Public Sociology series

Series Editors: **John Brewer**, Queen's University, Belfast Northern Ireland and **Neil McLaughlin**, McMaster University, Canada

---

The *Public Sociology* series addresses not only what sociologists do, but what sociology is for, and focuses on the commitment to materially improving people's lives through understanding of the social condition. It showcases the wide diversity of sociological research that addresses the many global challenges that threaten the future of humankind.

Forthcoming in the series:

Critical Engagement with Public Sociology
A Perspective from the Global South
Edited by **Andries Bezuidenhout**, **Sonwabile Mnwana** and **Karl von Holdt**

Out now in the series:

Erich Fromm and Global Public Sociology
Neil McLaughlin

The Public and Their Platforms
Public Sociology in an Era of Social Media
Mark Carrigan and **Fatsis Lambros**

Public Sociology As Educational Practice
Challenges, Dialogues and Counter-Publics
Edited by **Eurig Scandrett**

Find out more at

bristoluniversitypress.co.uk/public-sociology

Public Sociology series

Series Editors: **John Brewer**, Queen's University, Belfast, Northern Ireland and **Neil McLaughlin**, McMaster University, Canada

---

International editorial advisory board:

Find out more at

bristoluniversitypress.co.uk/public-sociology

A PUBLIC SOCIOLOGY OF WASTE

Myra J. Hird

BRISTOL
UNIVERSITY
PRESS

First published in Great Britain in 2022 by

Bristol University Press
University of Bristol
1–9 Old Park Hill
Bristol
BS2 8BB
UK
t: +44 (0)117 954 5940
e: bup-info@bristol.ac.uk

Details of international sales and distribution partners are available at bristoluniversitypress.co.uk

British Library Cataloguing in Publication Data
A catalogue record for this book is available from the British Library

ISBN 978-1-5292-0655-5 hardcover
ISBN 978-1-5292-0659-3 ePub
ISBN 978-1-5292-0658-6 ePdf

The right of Myra J. Hird to be identified as author of this work has been asserted by her in accordance with the Copyright, Designs and Patents Act 1988.

Cover design: Andrew Corbett
Front cover image: Andrew Corbett
Bristol University Press uses environmentally responsible print partners
Printed and bound in Great Britain by CPI Group (UK) Ltd, Croydon, CR0 4YY

FSC
www.fsc.org
MIX
Paper from
responsible sources
FSC® C013604

Contents

List of Figures

Acknowledgements

It is always the case that many people provide the cognitive and emotional sustenance that I need to write. This is especially the case for *A Public Sociology of Waste*, as I wrote much of it during the COVID-19 pandemic, bunkered down with my family during a long lockdown. First of all, I thank John D. Brewer for inviting me to contribute to this book series. I also thank my editor, Shannon Kneis, for her unwavering support and flexibility, especially through the (continuing) pandemic. I thank the Social Sciences and Humanities Research Council of Canada for funding for some of the case studies described in this book. An earlier version of Chapter 3 appears in *Canada's Waste Flows* (2021, McGill-Queen's University Press) and is used with permission. An earlier version of Chapter 5 appears in Hird, M.J. and Riha, J. 'Prepping for the [insert here] apocalypse and wasting the future', in Z. Gille and J. Lepawsky (eds) *Handbook of Waste Studies*, New York and London: Routledge, 2021, pp 305–21 and is used with permission. I wrote Chapter 4 with Jacob Riha's assistance. An earlier version of the placental research in the Appendix appears in Yoshizawa, R. and Hird, M.J. (2019) 'Schrödinger's placenta: determining placentas as (not)waste', *Environment and Planning E: Nature and Space* 3(1): 1–17 and is used with permission. Figure 3.2 'Recycled content, different material' is reprinted with permission from Vendries, J., Sauer, B., Hawkins, T.R., Allaway, D., Canepa, P., Rivin, J. and Mistry, M. (2020) 'The significance of environmental attributes as indicators of the life cycle environmental impacts of packaging and food service ware', *Environmental Science & Technology*, 54: 5356–64. Figure 3.3, 'The growing gap between available metals and minerals and current and growing demand'. Figure 3.4, 'Per capita waste generation and diversion rates in Ontario' is reprinted with permission from Scott Lougheed. Figure 4.2, 'Global fossil fuel consumption'. Figure 6.1, '*Anamnèse 1+1*, Alain-Martin Richard. Frédéric-Back Park', is reprinted with the artist's permission.

I tend to think of research as walking along a winding path, often with few signposts, and so I very much appreciated the Zoom meetings that I regularly had with my graduate students during this time. I thank Jacob Riha for his work on plastics waste and the prepping phenomenon. I am also grateful to Hillary Predko, who provided excellent copyediting assistance, and to Micky Renders, Aja Rowden and Gabrielle Dee. Scholars whose work I greatly admire – Kathryn Yusoff, Romain Garcier, Laurence Rocher, Judy Haschenburger, Sabrina Perić, Peter van Wyck, Tora Holmberg and Kyla Tienhaara – have all walked with me during the course of this research, making it entirely more enjoyable. And I thank Christophe Merle. I dedicate this book to Inis and Eshe who have let me share their paths.

Series Editors' Preface

Sociology is a highly reflexive subject. All scholarly disciplines examine themselves reflexively in terms of theory and practice as they apply what the sociologist of science Robert Merton once called 'organised scepticism'. Sociology adds to this constant internal academic debate a vigorous, almost obsessive, concern about its very purpose and rationale. This attentiveness to founding principles shows itself in significant intellectual interest in the 'canon' of great thinkers and its history as a discipline, in vigorous debate about the boundaries of the discipline, and in considerable inventiveness in developing new areas and subfields of sociology. This fascination with the purpose and social organization of the discipline is also reflected in the debate about sociology's civic engagements and commitments, its level of activism and its moral and political purposes.

This echoes the contemporary discussion about the idea of public sociology. 'Public sociology' is a new phrase for a long-standing debate about the purpose of sociology that began with the discipline's origins. It is therefore no coincidence that students in the 21st century, when being introduced to sociology for the first time, wrestle with ideas formulated centuries before, for while social change has rendered some of these ideas redundant, particularly the social Darwinism of the 19th century and functionalism in the 1950s, familiarity with these earlier debates and frameworks is the lens into understanding the purpose, value and prospect of sociology as key thinkers conceived it in the past. The ideas may have changed but the moral purpose has not.

A contentious discipline is destined to argue continually about its past. Some see the roots of sociology grounded in medieval scholasticism, in 18th-century Scotland, with the Scottish Enlightenment's engagement with the social changes wrought by commercialism, in conservative reactions to the Enlightenment or in 19th-century encounters with the negative effects of industrialization and modernization. Contentious disciplines, however, are condemned to always live in their past if they do not also develop a vision for their future; a sense of purpose and a rationale that takes the discipline forward. Sociology has always been forward looking, offering an analysis and diagnosis of what C. Wright Mills liked to call the human condition.

Interest in the social condition and in its improvement and betterment for the majority of ordinary men and women has always been sociology's ultimate objective.

At the end of the second millennium, when public sociology was named by Michael Burawoy, there was a strong feeling in the discipline that the professionalization of the subject during the 20th century had come at the cost of its public engagement, its commitment to social justice and its reputation for activism. The vitality and creativity of the public sociology debate was largely fuelled by what Aldon Morris called 'liberation capital', created in social movements of political engagement outside of the universities in the years after the social turmoil and changes of the 1960s.

The discipline has mostly reacted positively to Burawoy's call for public sociology, although there has been spirited dissent from those concerned with sociology's scientific status. Public sociology represents a practical realignment of the discipline by encouraging a focus on substantive and theoretical topics that are important to the many publics with whom the discipline engages. Public sociology, however, is also a normative realignment of the discipline through its commitment to enhance understanding of the social condition so that the lives of people are materially improved. Public sociology not only changes what sociologists do; it redefines what sociology is for.

Sociology's concern with founding principles is both a strength and a weakness of the discipline. Nothing seems settled in sociology; the discipline does not obliterate past ideas by their absorption into new ones, as Robert Merton once put it, as the natural sciences insist on doing. The past remains a learning tool in sociology, and the history of sociology is contemporaneous as we stand on the shoulders of giants to learn from earlier generations of sociologists. We therefore revisit debates about the boundaries between sociology and its cognate disciplines, or debates about the relationship between individuals and society, or about the analytical categories of individuals, groups, communities and societies, or of the primacy of material conditions over symbolic ones, or of the place of politics, identity, culture, economics and the everyday in structuring and determining social life. The boundaries of sociology are porous, and as many sociologists have asserted, the discipline is a hybrid, drawing ideas eclectically from those subjects closely aligned to it.

This hybridity is also sociology's great strength. Sociology's openness facilitates inter-disciplinarity, it encourages innovation in the fields to which the sociological imagination is applied and opens up new topics about which sociological questions can be asked. Sociology thus exposes the hidden and the neglected to scrutiny. There is very little that cannot have sociological questions asked of it. The boundaries of sociology are thus ever expanding and widening; it is limitless in applying the sociological imagination. The tension between continuity and change – something evident in society generally – reflects thus also in the discipline itself. This gives sociology a frisson that is both fertile and fruitful as new ideas rub up against old ones and as the conceptual apparatus of sociology is simultaneously revisited and renewed. This tends to work against faddism in sociology, since nothing is entirely new, and the latest fashions have their pasts.

Public sociology is thus not itself new and it has its own history. Burawoy rightly emphasized the role of C. Wright Mills, and broader frameworks allow us to highlight the contribution of the radical W.E.B. DuBois, the early feminist and peace campaigner Jane Addams, and scores of feminist, socialist and anti-racist scholars from the Global South, such as Fernando Henrique Cardoso in Brazil and Fatima Meer in South Africa. Going back further into the history of public sociology, the Scots in the 18th century were public sociologists in their way, allowing us to see that Burawoy's refocusing of sociology's research agenda and its normative realignment is the latest expression of a long-standing concern. The signal achievement of Burawoy's injunction was to mobilize the profession to reflect again on its founding principles and to take the discipline forward to engage with the relevance of sociology to the social and human condition in the 21st century.

Despite the popularity of the idea of public sociology, and the widespread use of such discourse, no book series is singularly dedicated to it. The purpose of this series is to draw together some of the best sociological research that carries the imprimatur of 'public sociology', done inside the academy by senior figures and early career researchers, as well as outside it by practitioners, policy analysts and independent researchers seeking to apply sociological research in real-world settings.

The reflexivity of professional sociologists as they ponder the usefulness of sociology under neoliberalism and late modern cosmopolitanism will be addressed in this series, as the series publishes works that engage from a sociological perspective with the fundamental global challenges that threaten the very future of humankind. The relevance of sociology will be highlighted in works that address these challenges as they feature in global social changes but also as they are mediated in local and regional communities and settings. The series will thus feature titles that work at a global level of abstraction as well as studies that are micro ethnographic depictions of global processes as they affect local communities. The focus of the series is thus on what Michael Ignatieff refers to as 'the ordinary virtues' of everyday life, social justice, equality of opportunity, fairness, tolerance, respect, trust and respect, and how the organization and structure of society – at a general level or in local neighbourhoods – inhibits or promotes these virtues and practices. The series will expose through detailed sociological analysis both the dynamics of social suffering and celebrate the hopes of social emancipation.

The discourse of public sociology has permeated outside the discipline of sociology, as other subjects take up its challenge and re-orientate themselves, such as public anthropology, public political science and public international relations. In pioneering the engagement with its different publics, sociology has therefore once again led the way, and this series is designed to take the debate about public sociology and its practices in new directions. In being the first of its kind, this book series will showcase how the discipline of sociology has used the language and ideas of public sociology to change what it does and what it is for. This series will address not only what sociologists do but also sociology's focus on the commitment to enhance understanding of the social condition so that the lives of ordinary people are materially improved. It will showcase the wide diversity of sociological research that addresses the many global challenges that threaten the future of humankind in the 21st century.

Few sociologists have done as much to highlight global threats to human well-being and survival as Professor Myra J. Hird, and thus this book on the public sociology of waste is a valuable contribution to our series. An interdisciplinary sociologist with a global reputation who provides national leadership in the social sciences in Canada where she is based, Professor Hird has written extensively about sexualities, violence, science and evolution, the environment and here, the public sociology of waste. She is a polymath. This book offers a theoretical framework for thinking about the sociology and politics of waste that illuminates the problem, shatters a number of myths and suggests ways forward for scholars to engage the public on this essential problem that is all too often framed as private troubles solved by recycling and reduced consumption.

Like much of the best public sociology, this book started with conversations with neighbours, observations in parks and serious engagement with the existing literature in the field of the sociology of the environment. The result is a significant contribution to scholarship on waste that insists on a structural analysis to locate our understanding of the threat to human life on our planet represented by global climate change and degradation of our natural environment. The book is both rigorously theoretical and politically important in the ways Hird insists on carefully examining how our definitions of waste, and the ways the questions are framed in public debates, are deeply shaped and distorted by powerful global capitalist corporations, settler colonial states and their culture, and ideological cultural workers in advertising, mainstream media and public intellectual debate.

A further strength of the book is its biographical location in Canada, with the powerful support for many of the Indigenous perspectives on the environment that are being articulated so powerfully today by Indigenous activists, theorists and movements around the world as a challenge to capitalist settler colonialism.

Individualizing the solution to environmental waste is seen thus as ideological. Hird's book is a sustained argument for thinking about solutions for waste and saving oceans, green spaces and the climate, with both analysis and policy prescriptions, that puts the blame on extractive industries, corporate and political elites and the capitalist and settler-colonial nature of the world's most wealthy nations.

Hird's book adds much to this series on public sociology because of her highly visible presence talking about these environmental issues, among other policy questions, on Canadian media, but also because of the case she makes for a particular type of public environmental sociologist. The granting agencies in Canada call a lot of what Myra Hird does 'knowledge mobilization'. We prefer, however, to call it a form of interdisciplinary and deeply engaged public sociology. The book is, to be sure, scholarly as it marshals detailed empirical case material, theoretical sophistication and historical breadth. Yet *The Public Sociology of Waste* has a message that is a powerful case for good sense and political vision to avoid waste destroying our planet. Thus, it will resonate with broad publics and in our view admirably represents the best kind of public sociology.

John D. Brewer and Neil McLaughlin
Belfast, Northern Ireland, and Toronto, Ontario
December 2021

1

The Public Problem of Waste

We are nowhere near 'peak waste.'

Hoornweg et al, 2014: 117

When we read news about waste – plastic bags clogging water drains in India and causing contaminated drinking water, child labour used to dismantle used electronics in Malaysia, space junk orbiting earth or PPE masks washing up on our shores – our waste problem seems clear: waste is being mismanaged! The solution appears similarly obvious: we must better manage our waste! But the devil, as the famous idiom goes, is in the details. What we understand the problem to be – the mismanagement of waste – depends on how we actually define waste. Whose waste are we referring to? Who should be responsible for managing it better? And what, more specifically, would register as 'better' waste management?

At first glance, the definition of waste seems equally obvious. Waste is all of that stuff that we once wanted but no longer want (Strasser, 1999). Waste is all of those things we put in our trash can, and if our local waste services are functioning well, are whisked away from our homes to quickly become out-of-sight and out-of-mind. Yet, waste turns out to be a rather complex problem involving different rightsholders and stakeholders, temporalities, geographies, political economies, transnational agreements, regulations and policies, and cultural traditions that disproportionately affect a range of publics. To introduce this complexity, let's consider the following waste snapshots:

Snapshot 1

The Republic of the Marshall Islands is a United States associated state and comprises some 1,156 islands in the Pacific Ocean, north of New Zealand. With a total population of just over 58,000, most of its territory (over 97 per cent) is water. Beginning around the 10th century, successive waves of colonizers and settler colonizers claimed the islands, from Micronesians, to Spanish, to German, to Japanese, and finally to Americans during World War II. The US began nuclear bomb testing on the Marshall Islands' Bikini Atoll in 1946 and continued detonating nuclear arms for over a decade. During this period, the US exploded 23 nuclear weapons, first above-ground and then underground. The second – Baker test – detonation contaminated all of the surrounding ships, leading Glenn T. Seaborg, chair of the Atomic Energy Commission, to call it "the world's first nuclear disaster" (in Weisgall, 1994: ix). As well as widespread lasting nuclear contamination affecting generations of islanders, flora and fauna, this military operation also left 95 navy ships, including cruisers, destroyers, submarines, attack transports, landing ships, carriers and battleships, all carrying fuel and some with live ammunition. And it also left the Runit Dome, known locally as 'The Tomb', which is a bomb crater made from the 18kt Cactus detonation in 1958. As Peter van Wyck details:

> Between 1977 and 1980, contaminated topsoil and debris from the atoll (including 16,000 items of WWII ordinance, such as unexploded artillery projectiles, mortar shells, hand grenades, and small arms ammunition) were bulldozed into the crater, and a 45 cm concrete cap, or dome, was constructed on the surface. The Dome is now at risk of failure from deterioration, saltwater incursion, vulnerability to typhoons and sea level rise. (in Kavanagh, 2020: 49)

Since the first nuclear detonation on 15 July 1945, there have been some 2,056 nuclear detonations across the globe, most of them exploded on colonized Indigenous lands far away from the capitals of colonizing forces (Arms Control Association, 2020). France, for instance, tested its first nuclear weapon, code-named 'Gerboise Bleue' (Blue Desert Rat) in Algeria in February 1960, which France had violently invaded and then colonized. This first French nuclear test was recorded at 70kt, or as powerful as the US bomb dropped on Hiroshima, Japan, during World War II. France also made use of some of its other colonies in the French Polynesian atolls in the South Pacific as well as controversially conducting the last of its 210 nuclear tests there in 1996, during the Comprehensive Nuclear Test-Ban Treaty (CTBT) negotiations in Geneva. France signed on to the treaty only after international protests included French export boycotts. Only in 2009 did France's Senate acknowledge the impacts of its testing programme and provide some compensation to civilian and military veterans.

France's nuclear testing had carcinogenic and other negative health effects on local residents: atmospheric plutonium-239 concentrations were found to be four times greater in these French colonies than in continental France, leading in some cases to the evacuation of whole islands. It also resulted in considerable damage to the environment. Radiation from the blasts led to declines in livestocks and biodiversity. And, like the Runit Dome, France stored radioactive waste (including plastic bags, clothing, metal scrap and wood) on the north coast of the colonized Mururoa atoll, in an area covering about 30,000 square metres. When cyclones hit Mururoa in 1981, radioactive waste was washed into the lagoon, including the plutonium-laden bitumen that had been used to contain the plutonium-239.

Snapshot 2

In July 2015, Lebanon authorities closed a major landfill in the Naameh region of its capital, Beirut. The landfill was over-capacity, and the local government attempted to resolve the problem by contracting with a British waste management company to export the waste to Russia. The British company, however, failed to submit the paperwork on time, leaving the garbage to rot in what became known as the 'river of trash' (Hume and Tawfeeq, 2016: np). In response to the Bourj Hammoud/Jdeideh landfill reaching capacity, local authorities extended the landfill, and investigations by Human Rights Watch (2020) reported that waste burning – which has significant negative health and environmental consequences – continues to be used as a means of getting rid of the garbage. To date, officials have not resolved the mounting municipal solid waste (MSW) crisis in Beirut.

This is not the first time a municipality has attempted to export its waste problem. On 31 August 1986, the Liberian cargo ship *Khian Sea* was loaded with over 14,000 tons of incineration ash waste from Philadelphia, Pennsylvania. Exports, including waste, often change company hands several times between point of origin and final destination, and Joseph Paolino and Sons, the company contracted by Philadelphia, subcontracted the shipment to Amalgamated Shipping Corp and Coastal Carrier Inc, who intended to offload the waste shipment in the Bahamas. When the Bahamian government refused to allow the cargo ship to dock, it triggered a more than two-year saga. The subcontractors and the crew of the *Khian Sea* attempted to offload the waste in several countries, including Honduras, Panama, Bermuda, Guinea Bissau, the Dutch Antilles and the Dominican Republic (Leonard, 2010). All refused to accept the waste. The subcontractors even tried to return the waste to its origin, Philadelphia, where officials also refused to repatriate its incineration garbage. Then, in January 1988, the crew dumped some 4,000 tons of the toxic waste in Haiti, calling it 'topsoil fertilizer' (Reeves, 2001).

Greenpeace alerted the Haitian government to the illegal dumping, but the ship left before government officials could compel the crew to reload the waste. While some of the waste was sequestered in a bunker, most of it remained on the beach, open to the environmental elements. But the story does not end there: the crew attempted to offload the remainder of its load in Sri Lanka, Singapore, Morocco, Yugoslavia and Senegal, without success. The ship even changed its name from the *Khian Sea* to *Felicia* and then to *Pelicano* in an attempt to disguise its waste load. Finally, in 1988, over two years after the cargo ship left the United States, the ship's captain and crew dumped the ash waste in the Atlantic and Indian Oceans. In 1997, the New York City Trade Waste Commission agreed to give Eastern Environmental Services a New York operating licence on condition that it contribute to the cleanup in Haiti. The city of Philadelphia contributed $50,000 to the cleanup operation. In 2000, Waste Management Inc, one of the world's biggest multinational waste management corporations, shipped some 2,500 tons of the toxic ash and contaminated soil to Florida, where it remained on the ship for two years before being finally offloaded to a landfill in Pennsylvania. The long and very public saga of this orphaned waste shipment contributed to the initiation of the Basel Convention's transboundary waste export regulations (Leonard, 2010).

Snapshot 3

As early as the 1950s, sailors were reporting that turtles and other marine life were ingesting plastics (Hamilton and Feit, 2019). But it was in 1997, when oceanographer and sailing boat race enthusiast Charles Moore was returning from a Los Angeles-to-Hawaii race and he and his crew spotted what is now known as the North Pacific Gyre or the Pacific Trash Vortex (National Geographic, 2019), that news and social media really turned its attention to plastics ocean waste. In his 2009 TED Talk, Moore noted that the Gyre contained more plastic than plankton, on which feeds and therefore sustains the ocean's sea life. The North Pacific Gyre swirls in the ocean currents between Japan and the United States and is some 20 million kilometres in size. In 2016, Moore and his crew discovered another – bigger – gyre of garbage, called the South Pacific Gyre or Garbage Patch, that swirls between South America and Australia (Cirino, 2017).

There are an estimated 4.8 to 12.7 million tons of plastics dispersed in the planet's oceans (Jambeck et al, 2015). Most of these ocean plastics consist of small pellets. In almost all cases, plastics do not disappear but rather simply break down into smaller and smaller pieces until they become microplastics (4mm in diameter or less). As Moore noted when he first observed the North Pacific Gyre:

> Remember eating sprinkles on cupcakes as a kid? The tiny little colorful sugary beads. … So colorful and delicious and always associated with good times! … But, if you spilled them, man oh man, what a task it was to clean up all those hundreds of colorful tiny sprinkles. This thought has been running through my head over the last few days as we've been inspecting more and more trawls heavily laden with plastic. (Cirino, 2017: np)

Most (about 80 per cent) of the plastics in the ocean come from the land. The other 20 per cent consists mainly of plastics debris from boats. Ocean plastics have been found at all levels of the food chain, including the human digestive system (Hamilton and Feit, 2019). Attached to these trillions of plastics pieces are toxic chemicals such as Persistent Organic Pollutants that are harmful to human and inhuman life. Moreover, the breakdown of these plastics when exposed to light and other ambient conditions releases methane and other gases that significantly contribute to global warming (ibid, 2019). The Ellen MacArthur Foundation (2016) predicts there will be more plastics than fish in the oceans by 2050.

Introduction

These brief snapshots provide a window into the diversity of waste issues that we, as a global society, are facing. The United States' nuclear testing on Indian land in Nevada and New Mexico, for instance, has been so extensive that the US government has labelled these areas 'sacrifice zones' (Krupar, 2013; Masco, 2006). All nuclear waste is stored temporarily: only Finland claims to have found a permanent nuclear waste storage system. Nuclear radiotoxicity endures for upwards of 100,000 years, or 3,000 human generations (van Wyck, 2005). The uncertainties of nuclear waste disposal leave the implementation of technical designs mired in controversy (Durant and Johnson, 2009; Solomon et al, 2012). The Runit Dome is just one nuclear repository dotting our global landscape whose temporary status is vulnerable to climate and other environmental changes. There are different types of radioactive waste, with different effects on human and animal bodies and the environment. Moreover, since nuclear energy produces radioactive waste throughout the entire fuel cycle – that is, from the mining of uranium, to reactor decommissioning (Center for Sustainability, 2012; Nuclear Energy Agency, 2010), to reprocessing (that is, recycling) – spent fuel is surrounded by serious questions of proliferation and safety (Lagus, 2005; UNESCO International School of Science for Peace, 1998). Further, both nuclear waste repositories and landfills consume enormous amounts of energy derived from fossil fuels to sort, treat, store and transport waste (Chong and Hermreck, 2010). And warning future generations about these nuclear waste sites means 'saying something about a future twice as far from us as human written culture lies in the past – or roughly the entire span of time since the ice age ... [which] seems utterly impossible' (Galison, 2014). Nuclear waste, then, introduces crucial problematics concerning both waste's sequestration and its temporality.

The *Khian Sea* fiasco is by no means an isolated incident in an otherwise stable and equitable waste export system. In February 2013, 69 shipping containers of MSW labelled as 'recycling' travelled from Vancouver, Canada, to Manila, Philippines. In the ensuing political debacle in which Canada's federal government was forced through international humiliation and pressure to repatriate its garbage, Vancouver's waste travelled some 20,072 kilometres on trucks and ships using non-renewable fossil fuels, around the globe and back again, to be landfilled in Canada at a cost of $1.14 million. These cases illustrate that countries, and particularly countries in the globalized north such as France, the United Kingdom, Germany, Canada, the United States and Spain, export their waste problems. All kinds of waste is piling up in the globalized south, and local governments are finding that both waste volumes and increasingly toxic forms of waste are increasing faster than their diversion programmes are able to handle. The global circulation of waste is made even more complicated by the fact that the common view that waste always flows from rich to poor countries is erroneous. France is the largest legal net importer of hazardous wastes (O'Neill, 2000), and more electronic waste is exported and imported between non-OECD countries (such as the Philippines, Cambodia, Malaysia, Thailand) than from OECD countries (such as Canada, the United States, Britain and Germany) to non-OECD countries (Lepawsky, 2018). This complex and evolving set of international and national regulations and policies and a heterogenous mix of hazardous and non-hazardous materials means that we are, literally and figuratively, in a mess. And as Hoornweg observes in this chapter's epigraph, we are facing a future of more, not less, waste. Unless we frame the problem of waste differently.

Whether in the form of mining, nuclear, industrial, hazardous, sewage or municipal, and whether it is dumped, landfilled, incinerated, buried deep underground or processed into something else that eventually becomes trash, waste constitutes what will likely be the most abundant and enduring trace of the human for epochs to come (Crutzen and Stoermer, 2000). Countries as well as supra-national organizations such as the United Nations gather statistical information on the production, diversion and disposal of different types of waste, including MSW; industrial, commercial and institutional (ICI) waste, which may include medical wastes; construction, renovation and demolition waste; household hazardous and special wastes; organics; packaging and printed paper, including paper and plastics; agricultural wastes, including things like animal sewage lagoons; extraction wastes, including mining; and nuclear wastes. But beyond these apparently rather neat statistical categories lies a labyrinth of complexity. This book is an effort to unpack this complexity using public sociology as a tool.

The public sociology of waste

The term 'public sociology' was coined in 1988 by Herbert Gans, during his Presidential Address at the American Sociological Association (ASA) conference entitled 'Sociology in America: The Discipline and the Public'. Before this, sociologists had certainly emphasized in their research and in their teaching the critical need to engage with pressing public issues, and in ways that would interest, galvanize and empower members of the public to better understand the complexities of public problems and their role in resolving them. For instance, during the late 1800s at Atlanta University, W.E.B. Du Bois employed empirical sociology to challenge widely held social Darwinist theories of Black people's intellectual inferiority. He went on to write books such as *The Philadelphia Negro* (1899) that detailed the lives of Black people living in Philadelphia and argued strongly – again using empirical research – that a 'submerged tenth' (which later became known as the 'talented tenth') of the Black population had the capacity to lead Black people to greater emancipation (Lewis and Willis, 2005).

Contemporaneously, public sociology is most often associated with Michael Burawoy, who said this at his 2004 ASA Presidential address:

> As mirror and conscience of society, sociology must define, promote and inform public debate about deepening class and racial inequalities, new gender regimes, environmental degradation, market fundamentalism, state and non-state violence. I believe that the world needs public sociology – a sociology that transcends the academy – more than ever. Our potential publics are multiple, ranging from media audiences to policy makers, from silenced minorities to social movements. They are local, global, and national. As public sociology stimulates debate in all these contexts, it inspires and revitalizes our discipline. In return, theory and research give legitimacy, direction, and substance to public sociology. Teaching is equally central to public sociology: students are our first public for they carry sociology into all walks of life. Finally, the critical imagination, exposing the gap between what is and what could be, infuses values into public sociology to remind us that the world could be different.[1] (2004: np)

Central to my approach as an environmental sociologist is to examine the structures of society – not only institutional structures such as families, schools, legal systems, health care and so on but also race, gender, sexuality, religion, social class, nation, age and so on – that are empirically shown to structure people's lived experiences. *A Public Sociology of Waste* is concerned with how these factors structure people's lived experiences of waste. Waste is a global problem whose lived experience varies substantially depending on cultural history, nationality, social class, gender, race and so on. As John D. Brewer notes, the 'promise' with which C. Wright Mills oriented sociology 'was to cultivate an imagination in the social sciences that helped ordinary men and women grasp the intricate patterns of their own lives and to see how these connected with wider structural forces and processes about which they had no understanding and over which they had no control' (2013: 146). Throughout the book, I argue that public sociology may usefully provide tools for understanding and framing waste as a *socio-ethical issue* (as opposed to a techno-scientific or consumer-behavioural issue) that is only resolved through public dialogue focused on waste's impact on the 'distant, marginalized and strange other' (ibid: 158), and the long durée of waste's toxicity, which future generations will inherit. As such, the book is firmly 'problem oriented rather than discipline oriented' (ibid: 171) and draws upon a range of literature from sociology, Indigenous studies, anti-racist studies, geography, engineering, biology, anthropology, political studies and communication studies to re-orient waste as a global socio-ethical issue.

In order to examine the complexities of waste, I situate *A Public Sociology of Waste* as a critical foray into the ways in which waste is *framed* by key rightsholders and stakeholders and the consequences of these framings for public engagement with waste issues. As such, the central argument anchoring the book is this: how we understand the problem of waste and its solution(s) critically depends on how waste as a concept is framed. Understanding these frames opens a way for public sociology to assist in resisting these framings and to engage publics in creating their own framing.

Throughout this book, I argue that one of the major consequences of these different frames is that the world's waste has become a 'wicked problem'. Wicked problems are ones that are difficult to resolve because their multi-dimensional complexity means that solutions are often contradictory: solving one aspect of a wicked problem may well open up a different problem or problems. For instance, as Chapters 3 and 4 show, increasing plastics waste recycling increases our carbon emissions, which contributes to global warming. As Chapter 5 examines, protecting ourselves and our loved ones from waste in our environment often means increasing our waste footprint. And as Chapter 6 demonstrates, even if we drastically reduce our consumption (itself a contradiction in capitalist growth economies), we are left with a profound (in its scale and toxicity) waste legacy that will endure for an unknown number of future generations. Thus, waste is a wicked problem precisely because it presses publics to confront the environmental, political, economic, symbolic and cultural dimensions of contemporary global society.

Studying waste issues and the public sociology of waste in brief

I have been studying waste issues for years, poking around open garbage dumps, visiting energy-from-waste (EfW) facilities, state-of-the-art landfills and recycling hubs. I have observed people hovering over recycling bins and painstakingly trying to decide which one to toss their coffee cup lid in, as well as rather listlessly hovering over these bins myself. I have spoken with numerous engineers, waste operators, government officials, food bank volunteers, environmentalists, members of the public concerned about waste issues, and fellow waste researchers, and I have read government reports, scholarly articles, academic books and novels on the topic of waste. And with each insight I gain from these forays, the more questions I have, and the more fascinated by waste issues I become.

I am fortunate to have found, and been found by, a range of expert scientists, engineers, social scientists, humanities scholars and graduate students similarly interested in waste issues. Years ago, R. Kerry Rowe, civil engineer and world-leading landfill expert, first suggested that my interdisciplinary way of researching, and my interest in materiality (I had just published a book on microorganisms in which I developed a social theory concerned with the origins of sociable life; see Hird, 2009), might be harnessed to study waste issues. Before my conversations with Kerry, I had not thought in any serious way about waste (beyond bacterial metabolism), so I have Kerry to thank for introducing this topic to me, as well as for letting me visit his civil engineering laboratories and shadow some of his graduate students through their research projects. I happened to sit next to Peter van Wyck, communications scholar at Concordia University who researches nuclear waste issues, on a flight between Saskatoon and Edmonton: a chance encounter that led to our work partnership on a number of research projects that focus on waste issues. Allison Rutter, biologist, waste remediation expert and director of the Analytical Services Unit at Queen's University, graciously let me shadow some of her graduate students conducting research on waste in her laboratory, and – very significantly – introduced me to waste issues in Canada's Arctic. It is in these communities that I have most confronted, and been confronted with, my privilege and ignorance as a white settler Canadian (see Tuhiwai Smith, 2012; Simpson, 2014; and Todd, 2016). And it is this perpetual confrontation, and its discomfort, that prompted me to realize, early on, that a book about waste issues must do more than just acknowledge the strong association between waste and (settler) colonialism.

I am grateful to the Social Sciences and Humanities Research Council of Canada (SSHRC) for a number of research grants that have funded some of the case studies that illuminate the waste issues described in this book, and to other granting agencies such as the Institut des Sciences de l'Homme, the France Canada Research Fund, the Swedish Foundation for Strategic Environmental Research, Mistra and Formas for financially supporting my research and making the cross-national analyses featured in this book possible. I also secured ethics approval from Queen's University for all of the case studies recounted in this book. And I am particularly grateful to the graduate students with whom I have been fortunate to work on various waste issue case studies, and in particular Jacob Riha, Hillary Predko, Micky Renders, Aja Rowden and Gabrielle Dee. Much of my thinking about waste as a critical lens with which to reflect upon big, interconnected issues like race, gender, socio-economic privilege, global governance and environmental justice has benefitted from ongoing discussions with a generation of emerging scholars who I hope will carry the torch of waste studies.

As the saying goes, life is what happens while you are doing other things. Not long after I took on the project of writing this book, news reports started filtering into the Canadian news media about a strange new virus affecting people in Wuhan, China. We all know the rest of this story as it is continuing in real time for so many of us. The COVID-19 pandemic and its travel restrictions curtailed much of the empirical cross-national research that I had planned to complete for this book. As such, I make use of research that I was able to complete before travel restrictions made further field research impossible, mainly in Canada, but also in France, the United Kingdom, Sweden, New Caledonia and the United States.

After introducing some of the variety and scale of our global waste issues in the present chapter, Chapter 2, 'Framing Waste', is concerned with the major theoretical lens with which I build my rationale for a public sociology of waste. After examining how framing theory has been taken up within various studies and applied to various social issues, I examine the three major frames that organize our differential understandings of waste as both a problem and how it should be resolved. The first major frame is that of individual responsibilization, or as I call it, the *problem of amplification*. Within this frame, waste is limited to MSW, which is the waste that individuals, families and households produce and manage on a daily basis. That is, the *problem* of waste is framed by manufacturing and retail industries, as well as governments and (misled) members of the public as a problem of post-consumer waste. This frame masks the fact that, overwhelmingly and by orders of magnitude, almost all waste is produced before consumers enter the waste picture. Our global waste problem is produced by the extractive, manufacturing and retail industries that have deliberately amplified the consumer's role in waste production in order to deflect responsibility from themselves and on to consumers. This frame has proven to be highly successful, as Chapter 2 will examine.

The second primary frame is that waste is a resource, or as I call it *now-you-see-it, now-you-don't*. Through an examination of a seemingly mundane waste object – the human placenta – I demonstrate how experts conceptualize waste as rightly discardable material to the general public, but valuable to experts who transform waste into a resource. This frame is equally effective whether we consider human placentas or MSW that is incinerated to produce energy. The third frame gets us closer to the heart of a public sociology of waste. It concerns identifying waste, explicitly, as a *social justice issue*. Indigenous and other activist groups as well as large non-governmental organizations such as Greenpeace and Human Rights Watch, as well as an increasing number of waste studies researchers, are exposing the links between waste and economic, political and social injustice. While Indigenous and poor communities around the world have known of – and indeed live in close proximity to (if not on) – waste sites for generations, this reality has only more recently caught the attention of a wider public. Illuminating waste as a social justice issue highlights that dump site, landfill and incinerator siting, waste exporting and a host of other waste-related practices are differentially organized worldwide by powerful private corporate, industry and government interests. This examination highlights the need for a sociology of waste to reject the two predominant frames and to find ways to engage publics with waste as primarily a social justice issue.

Chapter 3, 'The Public Problem of Recycling', is the result of several years of research, some of which I undertook with two of my former graduate students, Scott Lougheed and Cassandra Kuyvenhoven. In this chapter, I take a deep-dive into the realities of recycling. Given the force and enthusiasm with which local and federal governments, communities and non-governmental organizations around the world have embraced, championed – and become dependent on – recycling as the method of decreasing waste and improving the environment, it is important to examine what recycling actually entails and achieves. That is, it is vital that we study the research on the material benefits and costs of recycling compared with other waste management methods, as well as the social, economic and political consequences of recycling. Only by considering recycling's environmental footprint as well as how recycling operates within our cultural, economic and political systems are we able to meaningfully get beyond recycling rhetoric that has, since the 1970s, forcefully and loudly declared that recycling is an environmental good. Of the 3Rs – reduce, reuse and recycle – recycling is the least environmentally friendly but does not disturb (on the contrary, it encourages) circuits of capitalist production and consumption. Chapter 3 argues that municipal governments in cooperation with industry successfully foster an 'environmental citizenship' identity based on individual and household waste recycling, even though this accounts for a tiny fraction of our waste production. In doing so, members of the public are encouraged to accept, endorse and reproduce the amplification and individual responsibility frame examined in Chapter 2. Within this frame, members of the public – and mainly women – are encouraged to survey and judge their own recycling behaviours as well as those of their neighbours, families and friends, rather than the much more voluminous quantities – and often greater toxicity – of industrial and military waste. As such, *recycling itself is the diversion*. This chapter argues that waste is an excellent example of how governments and industry implicitly cooperate to divert attention away from understanding waste as something that industries and governments produce and must take responsibility for and instead shift responsibility to members of the public to resolve.

Chapter 4, 'The Public Problem of Plastics', which I wrote with the assistance of Jacob Riha, continues the examination of waste as a public problem with a more specific focus on plastics. We have witnessed a monumental shift in public perceptions about plastics, from their creation as a 'miracle' technology that would advance the possibilities and comfort of our lifestyles in the 20th century, to their condemnation in the 21st century as a prolific and dangerous waste that is polluting our bodies and environment. This chapter concentrates on how oil and gas industries, from which plastics are derived, have structured public discourses about plastics in order to ensure their continual – and vastly expanding – production, despite national and international calls for plastics reduction. Using archival sources, this chapter explores how industry and governments frame plastics recycling as the viable solution to plastics waste. While China's plastics import ban and the Basel Convention's Transboundary Amendments suggest a heightening of restrictions on plastics production, this chapter argues that oil and gas industries are feverishly framing plastics recycling as a 'green' solution. But rather than this leading to a reduction in plastics production, as would be the case in a closed-loop or circular economy, plastics production is *increasing* due to plastics recycling. While the public is distracted by protracted debates about whether plastic straws should be replaced with paper or metal straws, the oil and gas industry is heavily investing in plastics recycling infrastructure, and exponentially increasing their oil, gas and plastics production, which is significantly contributing to climate change.

Chapter 5, 'The Public Problem of PPE Waste and Being Prepared', which I wrote with Jacob Riha, takes advantage of what has surely been a difficult, and for millions, devastating time of living through the COVID-19 global pandemic. Environmental tipping points, political instability, dwindling primary resources and global capitalist growth economies furnish myriad imagined apocalyptic futures. While the Intergovernmental Panel on Climate Change (IPCC) and other supra-governmental bodies focus on developing systems of resilience and adaptation for vulnerable populations, a geographically disparate group of people are mobilizing as 'preppers'. Once represented by social media and some scholars as individuals and small communities on the margins of mainstream society, the global COVID-19 pandemic is positioning prepping as a rational and, indeed, responsible response to disaster. This chapter focuses on prepping as a particular response to the uncertainty of our species' survival. Drawing on a range of theoretical traditions and empirical observations, it critically examines the various discourses and practices that preppers deploy in preparing themselves and loved ones for what they believe is the certainty of a survivalist future. Far from the experiences of millions of people who are forced into relentless adaptation due to unremitting poverty, inequality and global changes in climate, preppers largely plan for their imagined future by accumulating survivalist skills and *things*. That is, what most characterizes preppers is their mass consumption: preppers spend many thousands of dollars stockpiling resources like food, water and weaponry. Walmart, Costco and other mainstream consumer havens now offer emergency food storage 'kits', as publishing and distributing companies sell prepping guides for adults and children alike. Not only, then, do they eschew initiatives that seek to prevent an imagined future apocalypse, preppers influence the very conditions that they then say they are forced to respond to as they intensify the hegemony of over-consumption. As such, I argue that the increasingly popular phenomenon of prepping is a contemporary reiteration of western consumer/trashing culture, which feeds the global neoliberal capitalist system responsible for the very apocalyptic conditions preppers believe they are insulating themselves from. This chapter illustrates a real-time consequence of the individualization of responsibility – inverted quarantine. The COVID-19 pandemic is bringing into sharp relief what happens when we frame things as individual responsibility – those who can protect themselves do, and the rest are left in increasing vulnerability.

The final chapter, 'A Public Sociology of Waste', distils the major insights derived from the previous chapters. My central concern here is to explore how waste is framed by industry and governments (at all levels) to the public as something that human ingenuity and expertise has transformed from hazardous and harmful to something that is benign, and indeed, a 'green' energy resource. Crucially, this framing is about reassuring the public that neoliberal capitalism is compatible with environmental sustainability. I make the case that a public sociology of waste must advance the public's understanding of waste as primarily industry-produced and governed, and that sociologists and other waste studies scholars must – using theoretical and empirical tools – reframe waste as a global social justice issue.

2

Framing Waste

'Trash is our only growing resource.'

Crooks, 1993: 22

Introduction

Framing theory, or framing analysis as it is also called, is widely used by disciplines such as psychology, media studies, anthropology, political studies, social movement studies, rhetoric studies, history and economics to analyse how individuals perceive, make sense of, and communicate their understandings of reality. Within sociology, framing is understood, broadly, as a powerful means by which individuals make sense of the world around them, or as Robert Entman notes, framing is the means by which individuals 'select some aspects of a perceived reality and make them more salient in a communicating text, in such a way as to promote a particular problem definition, causal interpretation, moral evaluation, and/or treatment recommendation for the item described' (1993: 51). Erving Goffman, whose work has significantly influenced my research for decades, defined framing as a 'schemata of interpretation' (1974: 21). In *The Presentation of Self in Everyday Life* (1959) and *Frame Analysis: An Essay on the Organization of Experience* (1974), Goffman analysed framing as the process by which individuals and groups including communities 'locate, perceive, identify, and label' events around them such that these occurrences are given meaning and guide action (ibid: 21).

Media studies particularly uses framing theory to make sense of the ways that news media influences public discourse. News media build frames according to current societal norms and values and, more broadly, the cultural context, pressures from interest groups such as industries and corporations (who fund news media), political and ideological affiliations and so on. The difference in how news sources such as CNN, the *New York Times* and the *Washington Post* framed the 2021 Capitol Hill riots in Washington, DC compared with how Fox News, OneAmerica News Network and the Freedom First Press reported on this event illustrates how ideological and political affiliations coupled with interest-group pressures dramatically influence not only how events are framed but also how individuals and groups already cleaving to these affiliations differentially either readily accept or scrutinize these frames. That is, people are far more likely to accept a frame when it corresponds to their already-existing understanding of the world.

In what now appears to be an experiment foreshadowing real-life, in 1986 Amos Tversky and Daniel Kahneman conducted a well-known experiment known as the racialized 'Asian disease problem'. Two groups of volunteer participants were to:

> Imagine that the U.S. is preparing for the outbreak of an unusual Asian disease, which is expected to kill 600 people. Two alternative programs to combat the disease have been proposed. Assume the exact scientific estimate of the consequences of the programs are as follows:

> In a group of 600 people:
> Program A: 200 participants will be saved
> Program B: There is a ⅓ probability that 600 people will be saved, and a ⅔ probability that no people will be saved.

Seventy-two per cent of the participants in this group chose programme A, while 28 per cent of the participants chose programme B. The second group of participants, given the same scenario, were given the choice between:

> In a group of 600 people:
> Program C: 400 people will die
> Program D: There is a ⅓ probability that nobody will die, and a ⅔ probability that 600 people will die.

When this frame was used, 78 per cent of the participants chose programme D while only 22 per cent chose programme C.

Of course, programmes A and C are identical, as are programmes B and D. Tversky and Kahneman concluded that when the choice was framed as lives saved, participants chose the secure programmes (A and C). Conversely, when the choice was framed as lives lost/death, participants were far more likely to choose to gamble on probabilities (B and D). As Chapter 4 will detail, interest groups such as Plastics Europe and manufacturing companies such as Coca-Cola and ExxonMobil use framing theory to encourage members of the public to focus on the 'public goods' associated with their products such as versatility and consumer choice rather than the human health and environmental harm associated with their products. Environmental activist groups similarly recognize the importance of framing in, for instance, debates about whether to use the term 'climate change' or 'global warming' (Saul, 2018).

The public sociology of waste developed throughout this book centrally frames waste as a social justice issue. In most contexts, such as municipal government reports about increases in household waste and littering or radio and television news shows that air programmes devoted to the ever-increasing costs of waste disposal, waste is framed less as a socio-ethical issue and more as an issue requiring more and better technology coupled with individuals taking greater responsibility for their own waste as well as the waste of the world. Framing waste as a social justice issue would require open forms of democratic deliberation on such issues as our reliance on an economic system based on relentless growth and environmental depletion, the dependence of government on the extractive and manufacturing industries and our global system of over-production and over-consumption.

And democratic deliberation is only possible when we become cognizant and critical of the current dominant frames with which the problems of waste, and their solutions, are structured. The major stakeholders who frame waste issues are: (1) the extraction and manufacturing industries; (2) experts (who may or may not work for industry or government); (3) local, regional, national and international government officials; (4) non-governmental organizations and social media; (5) waste itself – that is, the materiality of waste; and (6) various publics who may or may not organize in communities. Each of these rightsholders and stakeholders frame waste issues such that the problems of waste and their solutions conform to parameters that cleave to and support the rightsholder and/or stakeholder's own interests. In short, framing is a means of achieving the aims of the framer. Throughout this book, I will argue that members of the public have been encouraged and cajoled, incited and lured, and – where and when necessary – manipulated, threatened and harmed into accepting and adhering to the frames of the dominant stakeholders: the extractive, manufacturing and retail industries and (to a lesser extent) government.

The problem of amplification: framing waste as an individual responsibility

As the chapters in this book detail, manufacturing and retail industries, local and national governments, numerous non-governmental organizations such as those attempting to advance 'zero waste' initiatives and many members of the public, frame waste as an individual problem whose solution is greater individual responsibility. Simply put, the frame goes like this: individuals and families produce our waste problem, and individuals and families must resolve this problem through increased and better recycling behaviours. Indeed, the assumption that the world's waste problem is caused by consumers predominates discourses about waste to the degree that it appears as a 'common sense' truth beyond scrutiny. This frame both amplifies the problem of MSW and also amplifies the responsibility that individuals, families and households bear in creating this waste, as it minimizes the waste created by the extractive, manufacturing and retail industries and the little responsibility that industry continues to shoulder for reducing this waste.

I illustrate this particular frame by drawing upon data from Canada because Canada is the world's leading producer of MSW (Conference Board of Canada, 2013). Between 1990 and 2005, Canada's per person MSW production increased by 24 per cent. By 2000, Canadians were producing more waste per person than Americans, and by 2005, Canada was generating 791 kilograms of MSW per person – well above the OECD 17-country average of 610 kilograms per person, and almost twice as much as Japan (ibid). This translates into 34 million tons of waste in a single calendar year. The latest available statistics report that Canadians produce annually an average of 777 kilograms of waste per capita (Statistics Canada, 2012). Even if Canada's overall recycling rate of 30 per cent is considered, Canadians still produce more than Japan's overall gross waste production (Ghoreishi, 2013: 8).

Waste management, as we mainly encounter it, almost exclusively concerns dealing with waste once a product has been consumed. In other words, the waste we almost exclusively focus on is: (1) the waste produced *post-consumption*, and; (2) the waste produced by individuals and households. For instance, managing e-waste focuses on post-consumer e-waste – those iPhones, Samsungs, iPads, laptops and desktop computers we no longer want – rather than the waste that goes into producing these and other electronic objects. But the reality is that industries extract and consume an extraordinary array and volume of materials in fabricating these products for our consumption. In his game-changing book *Reassembling Rubbish: Worlding Electronic Waste* (2018), Josh Lepawsky describes how some 90 million pounds of e-waste is recycled through Dell's Reconnect programme. This sounds like a lot until we learn that a single smelter operation in Mexico that produces copper and other metals used in electronics production produces 819,000 million tons of sulphuric acid waste. Lepawsky observes that the waste produced in extracting some of the materials used in electronics *from one smelter* is 1.8 times larger than e-waste exports *from the entire* United States.

In his analysis of the mining industry's 'fourth technological revolution', Martín Arboleda (2020) provides similarly illuminating data on the extractive industry's production of waste:

In the mining industry, the synergistic effect of innovations in robotics, biotechnology, artificial intelligence, and geospatial information systems, has exerted a fundamental overhaul in the extraction and processing of minerals. The implementation of such technologies, according to official figures, has allowed Latin American countries to achieve substantial increases in mineral exports relative to total exports – going, for example, from 39.7 percent in 2001 to 62.4 in 2010 in Chile; from 46 percent in 2000 to 65 percent in 2010 in Colombia; and from 45 percent to 61 percent for the same period in Peru. The economic profitability brought about by the smart and robotized mine, however, pales in comparison to its material footprint, as an average large-scale extraction site produces up to a thousand times more solid waste than those working with older technologies. To put this future into perspective, *a large open-cast mine can produce up to forty times more solid waste in one year than any Latin American megacity produces during the same period.* (2020: 11–12, emphasis added)

Arboleda goes on to point out that this already catastrophic waste-production is being compounded by the technology enabling mining companies to revisit defunct mines to extract low-grade mineral deposits:

As a result, mineral deposits that had not been mined because they were 'uneconomical' under older technologies are now being reopened and transformed into large mines across every corner of the world, putting enormous pressure upon water sources, livelihoods, and communities. The ability to mine low-grade mineral deposits has made the mining industry more profitable and *has increased the material footprint of mineral extraction by a factor of around 1,000 (in terms of the ratio of solid waste produced per gram of mineral extracted).* (2020: 49, emphasis added)

Here we confront one of the little-known realities of our global waste problem: most waste is generated at the extraction and production stages rather than at the consumer stage. This fact should profoundly shift our thinking about waste. Staying with Canada as our illustration, the 2012 edition of Statistics Canada's 'Human activity and the environment' publication (which is the most recent statistical data source for countrywide waste production) reports that the oil sands industry produced 645 million tons of waste (this is actually a 2008 statistic), the mining industry produced 217 million tons of mining waste and 256 tons of mine waste rock, the agricultural industry produced 181 million tons of livestock manure (this is a 2006 statistic), and households produced 34 million tons of MSW (a 2008 statistic). This means that industry (only including oil sands, mining and agriculture) produced 97.4 per cent of our total waste. Households produced 2.6 per cent of our total waste.

Now, if we figure in construction, demolition and institutional waste (which includes places like universities, schools and hospitals), then Canadian households produce significantly less than 2.6 per cent of Canada's total waste. Now let's factor in military waste. It is difficult to obtain statistics on how much waste the Canadian military produces (on Canadian territory, let alone abroad). We have some information about historical waste sites, and the Canadian military publishes some data on how much waste is produced on military bases, but again this is institutional waste (food and so on) rather than the waste produced through military operations such as munitions testing. To include military waste in the equation, the 2.6 per cent figure needs to again be reduced considerably. So, we may reasonably estimate that MSW accounts for around 1 per cent (or less) of Canada's total waste production. Let us also further consider Max Liboiron's (2013) important insight, that most municipal waste should actually be considered industrial waste that has been externalized to consumers in the form of packaging and disposable and poor-quality goods that individual households are then held responsible for sorting and recycling with unpaid labour, for waste and recycling companies to then get back from municipalities, paid for through public taxes.

The latest 'Environmental indicators report' published by Environment and Climate Change Canada (2018) does not disaggregate waste in the way that previous reports have. Under the banner 'socio-economic indicators', the 'solid waste disposal and diversion' category refers only to residential and non-residential MSW. These indicators, then, do not refer to industrial and/or military waste, nor municipal liquid waste. Nevertheless, the results point towards the same general findings: from 2002 to 2016, the total amount of waste collected in Canada increased by 11 per cent, or 3.5 million tons; and at least in 2016, the non-residential sector produced 59 per cent of disposed waste (Environment and Climate Change Canada, 2018). These data are sourced from municipal governments, which only collect data on local businesses and households. Again, if we were to include industrial and military waste, this figure would be considerably higher. Other reports, such as Deloitte's bombshell report, 'Economic study of the Canadian plastic industry, markets and waste: summary report to Environment and Climate Change Canada' (2019), have found that 87 per cent of plastics end up landfilled, with only around 9 per cent of plastics being recycled (recycling, as Chapter 3 will detail, is not a sustainable response to waste) (Deloitte, 2019: ii). The Deloitte report also states that packaging, construction and automotive waste accounts for 69 per cent of plastics waste. Silpa Kaza et al (2018) found that the amount of (just) industrial waste produced per year is 18 times greater than MSW.

Thus, manufacturing industries, waste management industries (who profit from MSW management) and governments (especially at the municipal level) purposefully frame waste discourse in ways that almost exclusively focus the public's attention on post-consumption waste (that is, municipal solid, and to a lesser extent liquid, waste) and individuals' responsibility for this waste. As Chapter 3 explores in detail, most people assume that our waste problems stem from individual consumption and not enough recycling, and the reason most people assume this is because our participation in discussions of waste is governed in specific ways that begin and end with individual responsibility and better technological innovations (better landfills, better EfW facilities and so on). In this way, as Liboiron points out, 'the individual rather than government or industry is represented as the primary unit of social change' (2010: 1).

Waste is largely framed as something that individuals and households produce because our global economy is structured by neoliberal capitalism, which emphasizes a market economy, enhanced privatization, government protectionism of coal, oil and gas industries, and a general entrepreneurial approach to profit maximization (Crooks, 1993; Foote and Mazzolini, 2012; for general discussions of governmentality, see Burchell et al, 1991; Foucault, 1984, 1988). The emphasis on individual responsibility operates within this capitalist rationale to manage waste in ways that do not disturb ever-increasing circuits of mass production and consumption (and therefore industry profit), producing an almost exclusive orientation towards the type of waste that individuals/households produce post-consumption and then disposal and diversion (diversion, as Chapter 3 details, is very largely limited to recycling, which – like disposal – maximizes industry profit (see Hawkins, 2006; Kollikkathara et al, 2009; Lynas, 2012). Globally, waste management is extremely profitable. Just dealing with MSW went from $205 billion in 2010 to a projected $375 billion by 2025 (Wilson and Velis, 2015).

The Keep America Beautiful 'Crying Indian' advertising campaign in the United States is a good illustration of the manufacturing industry purposefully framing waste as an individual problem and thereby deflecting attention away from their responsibility for creating and increasing waste and its human health and environmental harm. In the 1950s, beverage companies such as Coca-Cola were moving from glass to plastic bottles. Plastic bottles were easier and cheaper to manufacture, thereby increasing company profit. Companies using glass bottles were resistant to the bottle deposit system that required companies to buy-back the bottles, clean, and redistribute them because this resulted in less profit (MacBride, 2012; Rogers, 2006). In 1953, the state of Vermont tried to institute a compulsory bottle deposit that consumers would pay when they purchased disposable bottles and a ban on selling beer in non-refillable bottles. A consortium of American beverage companies, including the American Can Company, Continental Can Company, the US Brewers Foundation and the Owens-Illinois Glass Company quickly responded by introducing the Keep America Beautiful campaign, which focused on the problem of highway litter. Government agencies such as the Connecticut State Highway Department and the New York State Department of Public Works, as well as non-profit organizations such as the Izaak Walton League of America, quickly joined in the deflection effort. Together, industry and government organized and paid for brochures that were widely distributed via post to people's homes as well as several public service announcements designed to educate the public about the problem of public littering. In 1971, coinciding with the world's second Earth Day, the Keep America Beautiful campaign launched a television advertisement with the tag-line 'People Start Pollution. People Can Stop It', which became known as the 'Crying Indian' advertisement. The advertisement features an American Native Indian, Iron Eyes Cody, paddling a birch-bark canoe down a river. As he paddles his canoe to shore and steps out of the canoe, a white woman throws trash out of her car, which lands on Iron Eyes Cody's feet. As Cody looks at the trash, a single tear falls down his cheek. At this time of writing, this advertisement is still available to watch on YouTube.

The advertisement is an excellent illustration of industry and government successfully framing waste as a post-consumer (litter) problem created by individuals (members of the public discarding their individual waste on the roadside) and away from pre-consumer waste produced by manufacturing and retail industries whose profits increase when they are not regulated to provide, and pay for, bottle-refilling (reuse) services. Iron Eyes Cody turned out to not actually be an American Indian but rather an Italian American actor. But this seems to have been an insignificant truth against a persuasive advertising campaign that offered a simple message contrasting nature's purity against people's carelessness and disregard for the environment (for more details, listen to NPR's podcast, The Litter Myth, 2019). The success of the Keep America Beautiful campaign in framing the public as the problem rather than industry eventually led the tobacco industry to fund campaigns that carefully framed the problem away from the human health and environmental risks of smoking (which would lead to a decrease in tobacco industry profits) and towards the apparent problem of cigarette-butt littering. And recent research has demonstrated that Exxon, Mobil and ExxonMobil used this same framing strategy to publicly amplify consumers' contribution to climate change and deflect attention away from their own, much greater, oil and gas industry contribution, as well as using paid editorial-style advertisements to undermine climate change science (and scientists) while acknowledging the climate change reality in their internal documents (Supran and Oreskes, 2020). The Crying Indian and other Keep America Beautiful campaigns are regarded as the first highly successful example of corporate greenwashing.

Framing waste as resource: now you see it, now you don't

In tandem with the waste-as-individual-problem frame, extractive and manufacturing industries, experts and governments are also increasingly framing waste as a resource. In order to understand how this particular framing works, I turn to an empirical study of human placentas. At first glance, placental waste may appear tangential to our understanding of the public sociology of global waste issues. But it is precisely in closely examining this mundane object that, I argue, we gain critical insight into the often-subtle mechanisms through which stakeholders frame waste as a resource. My interest in human placentas as a waste object began as the result of a casual conversation I had with a colleague in Obstetrics and Gynaecology at the University of Ottawa, with whom I was conducting research on informed consent issues in health care. During our conversation, he mentioned that human placentas were transported from Kingston, Ontario, to Toronto (about three hours' drive away) and then stored in refrigerated facilities for later research purposes. This off-hand comment fascinated me, so much so that I designed a PhD project that one of my graduate students, Rebecca Scott Yoshizawa, completed in 2014 (see Yoshizawa, 2014).

An estimated 50 million kilograms of human placental material is produced worldwide every year. Depending on their meaning and significance within any given culture, human placentas are used in scientific research concerned with foetal and women's health, immunology and cancer; buried by a family or community (Lemon, 2002); flushed into the sewer system after being ground up at a hospital using a garburator-like machine (Callaghan, 2007); consumed by mothers or kin as food or medicine (see Beacock, 2012; Haraway, 2008; Ober, 1979; Young and Benyshek, 2010); melded with body surfaces as ingredients in cosmetic products such as hair serums and massage oils (Muralidhar and Panda, 1999); or used to train forensic detection dogs to recognize the scent of human remains (Hoffman et al, 2009; Judah, 2008). From 2010 to 2014, Rebecca and I conducted an empirical study that included semi-structured interviews with, and observations of, placenta scientists, as well as ethnographic participant observation in four laboratories and two hospitals in North America and Oceania. A total of 31 participants (19 women and 12 men) from ten different countries, including key informants with various levels of experience, investment and interest in placenta science, participated in this study. Until 2018, Rebecca continued to make observations at symposia and collected personal communications, following up on themes that emerged in our empirical study.

Our research uncovers the specific practices that placenta scientists use to disentangle placentas from material, affective, symbolic and moral attachments that might problematize their use in scientific experimentation. We argue that the process of defining placentas as not/waste entails a four-phase rationale that scientists use to effect these (re-)stabilizations, which begins with the claim that patients/mothers – and indeed nature itself – define placentas as waste. This is an important step: patients, hospital personnel and the public at large must concede that placentas are waste, and scientists employ various techniques to obviate other meanings that may be attributed to placentas. From here, a second phase involves scientists invoking what waste studies scholars call the 'waste hierarchy' that values waste as a resource (that is, extracting further value from a material's original use[s]). A third, subsequent, rationale identifies expert placental scientists as uniquely qualified to recognize placentas as waste-as-resource. A fourth and final phase involves the rationalization of placental scientists as motivated not only by professional and practical considerations but by a *moral imperative* to use placentas in scientific research. Our analysis indicates that the purported waste-ness of placentas potentiates their amenability to scientific experimentation and is foundational to scientists' claims about their moral relationship with broader publics. As such, we argue that the use of placentas for scientific research entails a paradox: scientists must define placentas as waste, and rely on women who have just birthed and their families to also do so, at the same time that they avow the value and utility of placental material. In so doing, scientists draw attention to the instability of the concept of waste: placentas must be stabilized as waste in order to be subsequently de-stabilized as waste, then re-stabilized as valuable scientific material, then subsequently de-de-stabilized as valuable scientific material in order to be finally re-re-stabilized as waste. I provide further details about this study in the Appendix.

This phased process of defining waste as a resource is by no means confined to placentas, or human or animal biological material for that matter. It is, indeed, the process by which waste, *in general*, is framed by the waste management industry and its experts (as Chapter 3 will examine) and the oil and gas manufacturing sector (as Chapter 4 will detail). Both of these industry sectors increasingly define waste as an energy source using the same step-wise process that Rebecca and I identified within the placental research community. In her original research, Laurence Rocher (2020) analyses how waste is increasingly framed as a valuable energy-producing resource, or as Rocher puts it, 'waste has become an energy issue' (2020: 98). And in a process very similar to what Rebecca and I found with regard to placental science, plastics waste is first defined as something that is useless to the public in order to be collected by municipalities (with the costs externalized to taxpayers) for it to then be redefined as an energy source and a public good (ibid). This is how, for instance, Sweden is able to boast that it has resolved its waste problem. Headlines such as 'Sweden's strange problem: not enough trash' (Murphy, 2017) present Sweden as a country of almost 'zero trash'. The Swedish government's own 'The Swedish Recycling Revolution' (Hinde, 2020) claim is based entirely on the reframing of waste as energy (in Sweden's case, electricity used for heating). What these remarkable claims – and this framing of waste as an energy resource – do not detail is the environmental costs of waste-to-energy incineration facilities (which I detail in Chapter 3), including highly toxic waste generation and the significant environmental harm accrued by transporting waste from other countries to Sweden. In this chapter's opening epigraph, Crooks calls attention to the irony that while we are decreasing (sometimes exponentially) the Earth's natural resources, we are increasing (sometimes exponentially) our garbage generation – that some government officials and industries are trying to frame as an environmental resource. It also purposefully deflects attention away from the fact that it is a linear economy (that is, the energy is used only once). And it masks the oil and gas sector's heavy hand in *increasing* plastics production (which makes up most of our MSW) and the waste management industry's *charging people twice*: once to collect our waste and a second time in selling our waste back to us in the form of energy, without in any way taking steps to reduce waste, and further increasing their profits through tax breaks (see Liboiron 2013).

Framing waste as a social justice issue

As Chapter 6 examines in detail, a significant challenge to the thus-far hegemonic framing of waste as either an individual post-consumption responsibility or as a profitable resource is to frame waste as a social justice issue. Framing waste as a symptom of environmental racism (Hird, 2021) requires us to attend to macro- and micro-structures and practices that influence when, where and how, and by whom, our waste is managed. One of the salient ways in which waste is framed as a social justice issue is to reveal the relationship between waste and racism. Benjamin Chavis first used the term 'environmental racism' in 1982 to refer to the polychlorinated biphenyl (PCB) waste that the United Church of Christ Commission for Racial Justice (of which he was executive director at the time) was trying to draw attention to, and force the government to remediate the Warren County PCB landfill in North Carolina (Chavis and Lee 1987). This contaminated waste was disproportionately affecting a predominantly Black community situated near the landfill. Chavis defined environmental racism as:

> racial discrimination in environmental policy making, the enforcement of regulations and laws, the deliberate targeting of communities of color for toxic waste facilities, the official sanctioning of the life-threatening presence of poisons and pollutants in our communities, and the history of excluding people of color from leadership of the ecology movements. (Chavis and Lee, 1987: 3)

Framing waste as a symptom of environmental racism means understanding waste within the context of overarching upstream critical issues involving historical and ongoing settler and non-settler colonialism, poverty and racialized and gendered relations of inequality. Waste problems are far more likely to be found in or near racialized and lower-income communities because these communities are less likely to have access to, and influence within or over, larger government power structures that create and enforce waste regulations, nor municipal-level governments who choose where to site waste facilities such as landfills and incinerators. Racialized and lower-income communities, in any country, are far less likely to have members who are sufficiently income-secure to be able to volunteer their time to protest waste issues, as middle-class and wealthy communities do. And racialized and lower-income communities are, by definition, income-challenged, and are therefore more likely to agree to a waste facility near their community in exchange for money and/or other payments in kind such as the building of recreation centres, as well as promises of employment. And racialized and lower-income communities are far less likely to be able to afford the engineering and legal experts required to take industry and/or governments to court in order to force them to take legal and financial responsibility for contaminant remediation (for an example of how a largely white and relatively wealthy community was able to use its considerable financial and cultural capital to refuse the siting of a landfill near their community, see Forkert, 2017).

In some places, communities are not just close to waste facilities but are rather built on and in the waste of more affluent regions (Amegah and Jaakkola, 2016; Yang and Furedy, 1993; Mothiba et al, 2017; Parizeau, 2006). Mega slums around the globe are increasing in number and size: Bantar Gebang in Indonesia is a 28–40 million ton open dump that accepts some 230,000 tons of waste per year (ISWA, 2016); Jam Chakro in Pakistan covers some 500 acres and has around five million people living in its vicinity. According to a United Nations environment report, some two billion people have no waste collection services and at least three billion people cannot access waste disposal facilities (Wilson and Velis, 2015). A number of studies (for example Dias, 2016; Dias and Fernandez, 2013) explore the ways in which people (and especially women and children) live on, and survive by picking through, dumps, in what Mike Davis describes as our 'planet of slums', in his book of the same title (2007).

One of the major aims of framing waste as a social justice issue is to expose the relationship between the ever-increasing production and consumption demands of the world's privileged and the human health, environmental, labour, family and cultural consequences for the world's disadvantaged. As François Jarrige and Thomas Le Roux observe, '[t]he decoupling of sites of production – subjected to direct pollution – from sanitized sites of consumption rendered almost imperceptible to the public the increase in emissions of toxic substances in the environment' (2020: 237). For instance, according to the Ellen MacArthur Foundation, up to half of all the clothing donated to used clothing stores in developed countries ends up being exported to poorer countries (2017). These so-called 'donations' overwhelm these nations' domestic textiles and clothing industries, leading the East African Community (Tanzania, Burundi, Rwanda, Kenya, Uganda and South Sudan) to ban second-hand clothing imports as of 2019. This reality is also illustrated by the case of electronics – all of those smartphones and watches, laptops and desktops that we cannot imagine our lives without. It is difficult to appreciate the planetary scale of the mining resource extraction required to furnish the raw materials – including rare earth materials such as ruthenium and indium – required for our electronic devices. As the previous section detailed, colossal amounts of toxic waste are produced just in extracting the required material. Then there is the energy required to manufacture the electronics, and the waste this manufacturing process produces. Add to this the waste produced in shipping these electronics around the globe, mainly in the form of carbon emissions from container shipping. Then using these electronics requires energy:

> Though individual usage may seem trivial, a simple internet request on Google equals the consumption of a 12-watt light bulb for 2 hours; sending a 1-megabyte attachment to ten correspondents requires the energy needed to move a car 500 meters. In an hour, around the world, 10 billion emails are sent, which corresponds to 50 gigawatt hours, or 4,000 Paris–New York round-trips by plane. (Jarrige and Le Roux, 2020: 304)

The constant marketing of 'new and improved' electronics and their planned obsolescence results in increasing volumes of electronic waste, only a small fraction of which is either reused or recycled. But while the appetite for electronics is increasing, our supplies cannot keep pace. Even if we were somehow able to recycle all of the metals and minerals contained in electronics, the current demand for electronics has already outpaced the supply that recycled electronic waste could produce. As Chapter 3 will show, as electronics demand increases, the minimum gap to be filled by primary resource extraction simply increases as well. Documentaries such as *The E-Waste Tragedy* (Dannoritzer, 2014) and other exposés by Greenpeace (Weyler, 2019) and human rights organizations detail the health impacts on poor Black/Indigenous/People of Colour children, women and men as they disassemble these electronics; what David Naguib Pellow calls 'toxic colonialism' (2009). When discussions of waste are isolated from profound forms of inequality, then it is far easier to maintain our focus on individual choices and individual responsibility.

Framing waste as a public problem: resisting the techno-fix and reframing the amplification

As ordinary citizens, we are most familiar with the waste that we ourselves create post-consumption: MSW. This category of waste amounts to a very small fraction of the total amount of waste that we, as a global population, are producing. In order to frame waste as a public problem, we have the immediate challenge of scale, which I will revisit throughout the book. The scale problem is that the public (families, households) are held disproportionately responsible for our global waste problem despite post-consumption consumer waste accounting for the smallest proportion of global waste generation. Conversely, the extractive and manufacturing industries produce far more waste – and often this waste is highly toxic – in the pre-consumption stages of any given product's life, while taking little financial responsibility to either reduce or even remediate the waste they produce. Thus, if we are to make any headway in resolving the globe's waste crisis, we need to critically analyse, in order to create forms of resistance to, the hegemonic frame of individualization and amplification that industry and government currently control.

As well, a political, economic and cultural system dependent upon constantly increasing production and consumption, and on a stewardship approach to environmental issues, ensures that waste continues to be framed largely in terms of technological innovation, jurisdictions and diversion practices produced through education, surveillance, sanction and censure. As such, waste management either capitalizes on or is victim to (depending on one's politics) what the Frankfurt School called 'technological rationality', 'in which the instrumental logic of rationalisation through "technology" colonises every last aspect of modern life, including and especially, thinking itself' (Grajeda, 2005: 316). We see this almost every day, as politicians and industry representatives emphasize the need for technological innovations to solve our environmental crises – from global warming and fossil fuel dependence to species extinction and plastics recycling, and that individuals take responsibility for our global waste crisis. Thus, industry pushes for more and better landfills, incinerators and EfW facilities. Waste is just another neoliberal capitalist entrepreneurial opportunity.

Throughout this book, I argue that public sociology is well adapted to shape deliberations concerned with our 'waste maker' global society because it is uniquely positioned to move deliberations from downstream responses (for instance, more recycling, better landfill technology) to much more challenging upstream issues. This reframing of waste is *only* achievable by engaging with various publics in ways that illuminate the relations between waste and wider societal issues concerned with over-production and over-consumption, poverty, racism, sexism and social justice. As such, the book's main contribution to public sociology is to detail how publics currently engage, and are engaged by, waste but also how publics might transform the framing of waste problems to reflect the realities of waste production and management, and solutions that benefit society and our environment rather than neoliberal capitalism.

3

The Public Problem of Recycling

Introduction

Some 15 years ago, when I began studying waste issues, I was facing a long commute to and from work that involved waiting for a series of buses and the metro. As is often the case with commuters, we sometimes struck up conversations while waiting for our public transportation. After a number of these conversations, I noticed a trend: when asked about my work, and telling people that I research waste issues, I always got the same response. One hundred per cent of the time. The responses would go something like this: "I feel so bad about all of the garbage we have everywhere. I try to recycle as much as I can." Sometimes my temporary travelling companion would go on to say that she, he or they did not know which recycling bin to put a certain item in. Sometimes the person would add that they had a neighbour/friend/family member that did not recycle as well as they should. Mainly women detailed teaching their children how to recycle. Now, years later, I am honoured to be regularly asked to give public presentations to community groups about waste issues. And my talks are uniformly met with the same range of questions that centre on one theme: how do we recycle more and better?

As Chapter 2 details, local and national governments, manufacturing and retail industries, and many non-governmental organizations concentrate on recycling. For instance, numerous local grass-roots 'zero waste' organizations across North America, Europe and the Antipodes are working towards achieving zero-waste lifestyles, which often, if not mainly, involves increasing their members' recycling behaviours. This focus is based on a number of assumptions that I will challenge in this chapter. The first premise is that increasing recycling decreases waste. Simply put, the assumption is that if we divert objects from the trash can and into the recycling bin, then we are decreasing our overall waste production. The second, related, premise is that by recycling, we are significantly decreasing our waste's global environmental footprint. Put another way, the assumption here is that individuals and families produce our waste problem, and individuals and families can (and must) resolve this problem through increased and better recycling behaviours. Indeed, the assumption that the world's waste problem is caused by consumers dominates discourses about waste to the degree that it appears as a 'common sense' truth beyond challenge. As such, recycling has, to a large degree, been 'black-boxed' in the service of increasing individual responsibility for our global waste crisis. The third, related, premise is that recycling is good for the environment, and certainly better than landfilling or incinerating waste.

The aim of this chapter is to examine recycling as *the* key 'go-to' solution deployed by the problem of amplification frame, which almost exclusively focuses on consumers and their post-consumption waste. This frame draws attention away from the much larger (by volume, weight and toxicity) proportion of extractive, manufacturing and retail waste produced before consumers ever encounter the products we buy. This frame simultaneously attributes all power, and also all responsibility, to individuals, families and households. Individuals have largely both accepted and shouldered this responsibility, and are thus willingly, if naively, participating in an amplification and deflection created by the manufacturing industry and corroborated by governments. In doing so, recycling draws attention away from the real sources of the global production of waste, while at the same time defining (and consigning) our contribution to environmental concerns to our consumption behaviours as the most important, if not only, form of community or societal participation. Indeed, I will argue that recycling has become critical to privileged lifestyles, whereby individuals morally offset environmental (and social justice) costs by recycling, or as Michael Maniates acerbically questions, 'plant a tree, buy a bike, save the world?' (2002: 43). Not only, then, does recycling morally sanction consumption but it also masks the true environmental and social justice costs of extraction and production, thus causing further harm to individuals, families, communities and our environment. And, as Chapter 5 will also argue, while believing that they are making a meaningful contribution to protecting the environment, privileged consumers may well be less involved in community activism and wider social justice issues that are critical to actually resolving our waste problem.

Making consumers responsible for our global waste problem

How have the people who produce the least amount of waste come to shoulder the responsibility for the world's waste? To address this question, we need to understand recycling within the context of a range of ways of managing waste. Many people are familiar with what is known as the 'waste hierarchy' (see Figure 3.1). It organizes how we manage waste, from the

Figure 3.1: The waste hierarchy

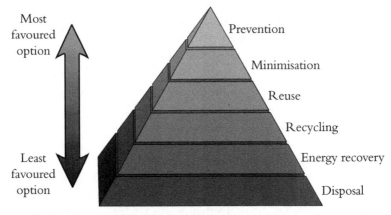

Source: Creative Commons, Wikipedia (CC BY-SA 3.0)

most to least preferable; that is, what is best to worst from an environmental perspective (Gharfalkar et al, 2015).

From the standpoint of protecting the environment, refusing or otherwise reducing or minimizing products is best. Then comes reusing products, and this may include refurbishment. Examples of this include using reusable coffee cups and 'pop-up' repair shops where consumers can learn to repair a toaster or other small appliance rather than buy a new one. Recycling and energy recovery appear at the base of the waste hierarchy, just above disposal (which refers to open dumping or landfilling waste). Recycling waste involves transforming a product from one thing (such as a plastic water bottle) into something else (such as a lawn chair). This transformation involves at least one, and usually several, mechanical and chemical processes, as we will see in the next section. Energy recovery (typically in the form of electricity or heat) is a particular form of recycling, whereby energy is created through the process of burning waste: waste is chemically transformed into energy. As an illustration, Sweden's claim to being 'zero waste' is based on their extensive use of EfW facilities throughout the country to provide electricity and/or heating to communities. Sweden actually imports waste from other countries (such as Finland and Germany) in order to have sufficient material to keep the EfW facilities operational (what civil engineers call 'feeding the beast'). It is important to note that energy recovery is near the base of the waste hierarchy because it is a linear system: the energy from burning waste is used only once and is not reused. As such, energy recovery is not part of the circular economy.

The waste hierarchy is a widely used aspirational system for individuals, organizations, industries and governments to govern their waste. In other words, the waste hierarchy is considered a valuable measuring tool in evaluating how we govern waste (European Commission, 2008; Hultman and Corvellec, 2012; Kurdve et al, 2015; van Ewijk and Stegemann, 2014). So, the more a community prevents waste (apex of the hierarchy) rather than disposes of its waste (bottom of the hierarchy), the better that community is managing its waste. Research shows that the waste hierarchy is frequently used by local, regional and federal governments as well as supra-national organizations such as the European Union to measure how effectively waste is being managed at local, regional, national and global levels.

It is vital to note that when waste studies researchers, grass-roots organizations, municipal governments and companies refer to the waste hierarchy, they almost exclusively refer to products *post-production*, that is, at the point at which consumers encounter products. Indeed, the waste hierarchy is mainly applied to MSW rather than, say, mining waste. As we saw in Chapter 1, post-consumption waste accounts for a small fraction of the total waste we produce. Indeed, in the 30-year update of the well-known 'The limits to growth' (2004) report, authors Donella Meadows, Jorgen Randers and Dennis Meadows describe the 'golden rule' that for every ton of waste produced by consumers (post-consumption), a whopping 20 tons of waste is produced in the extraction process alone. In the European Union, only 8 per cent of the 2.5 billion tons of waste produced in 2014 was household waste: 92 per cent of the waste was generated by the extractive and manufacturing industries (Cavé and Tastevin, 2021: 9). To take two everyday examples, 2.1 tons of materials are needed to manufacture a 79 kilogram washing machine in France, which is 27 times its weight. An 11 kilogram television requires 2.5 tons of materials, or 227 times the television's weight (ibid; Ademe, 2018: 186). Most of the difference between manufacturing the product and the product itself goes to waste.

That the waste hierarchy is almost always limited to post-consumption MSW is already revealing of an overarching system that favours maintaining/increasing production (extraction, manufacturing and retail) and consumption. Neoliberal capitalism favours an entrepreneurial approach to profit maximization with government intervention to supplement big corporate industries (such as oil and gas; see Chapter 4) through low or no taxation and other preferential arrangements. Thus, as we shall see, this overarching economic, political and social system favours attention to MSW for which individuals, families and households are held responsible because it is post-consumption waste. As such, it does not disturb ever-increasing circuits of mass production and consumption (and therefore industry profit).

In their article, 'Governing municipal waste: towards a new analytical framework' (2005), Harriet Bulkeley et al provide a thorough analysis of the use of the waste hierarchy in managing waste at local, regional, state and supra-national levels. They develop four 'modes of governance' that align with the waste hierarchy: disposal, diversion, eco-efficiency and waste as a resource. As waste's management has shifted in countries in the globalized north from individual 'mom and pop' operators to colossal multinational corporations such as Browning-Ferris Industries and Waste Management Incorporated (Crooks, 1993; Davies, 2008; Melosi, 2005), disposal has been – by far – the most pervasive mode of governance. As Bulkeley et al argue, 'with holes to fill and rubbish to get rid of, landfill has made almost incontrovertible (economic) sense as a waste management option in recent decades' (2005: 3). These companies and corporations increasingly control all aspects of the waste management system, from collection and haulage, to the landfills themselves. In countries like the United States, Canada, China and Russia, landfilling is a logical disposal tool because of the relatively enormous terrain available. However, this is not to say that far smaller countries such as France, the United Kingdom and Portugal do not also landfill waste. These geographically much smaller countries demonstrate a preference for incineration and waste export but they also continue to somewhat rely on landfills.

The second mode of waste governance that Bulkeley and her colleagues analyse is diversion, which is defined as moving items out of what would otherwise go into the waste stream (that is, disposal). Private waste management companies and corporations as well as local governments have overwhelmingly defined diversion as recycling. Regional, and sometimes national, levels of government often set diversion targets that local communities are expected to meet. Communities may well be monetarily rewarded if they meet these diversion targets or monetarily penalized if they fail to meet the targets. As well, these targets typically increase over time. Over a ten-year period, for instance, a region might be expected to increase its diversion rate by 20 per cent. And then in the next decade, this same region may be expected to increase its diversion rate by 60 per cent. With these, often ambitious, diversion targets, local communities are under pressure to find waste management options, and these options are mainly circumscribed to recycling as the 'go-to' solution.

To be clear, diversion has not surpassed disposal as the primary way of dealing with waste, although recycling companies and governments often present this as the case. Recycling is so popular with local governments because it overwhelmingly refers to post-consumption MSW, and as such, places responsibility on individuals, families and households and is paid for through taxation. In this way, local governments do not need to wade into the difficult issue of controlling the production of waste at the extraction or manufacturing stages of a product, which is currently beyond the remit of most local governments. The remit of local governments includes retail waste (which is usually placed in the category of Industrial, Commercial and Institutional) and post-consumption waste, and local governments may be reticent about attempting to regulate retailers with regard to their waste production lest this lead to retailers moving their businesses to other, less restrictive, areas. It has proven far easier for local governments to focus on the waste that consumers produce after they have consumed products. And not wanting to decrease retail business, local and regional governments have enthusiastically embraced recycling. Indeed, recycling has become emblematic of responsibly caring for the environment in general rather than a waste management technique near the least-preferable base of the waste hierarchy (Liboiron, 2010).

Bulkeley and her colleagues examine two further modes of governance: 'eco-efficiency' and 'waste-as-resource'. The authors note that, unlike disposal and diversion, eco-efficiency and waste-as-resource attempt to '[move] waste management options up the waste hierarchy' (2007: 2748) towards reuse rather than diversion (that is, recycling) and disposal. Already recognizing one facet of the environmental costs of recycling, Lily Pollans notes that these modes entail 'more dramatic and transformative action, including an emphasis on material reuse instead of more energy-intensive recycling' (2017: 2303). The eco-efficiency mode emphasizes governance partnerships between government and non-government stakeholders (and, we would expect, rights holders, although Bulkeley et al do not refer to rights holders specifically). These partnerships are meant to increase local governments' say in how waste is managed while also recognizing, and benefitting from, the 'specialist expertise' of private industry (Bulkeley et al, 2007: 2748). So, governments within this mode might well use financial incentives that 'redefin[e] the rationality of municipal waste management away from focusing on targets to achieving more concrete, local, social, and environmental objectives' (ibid). This eco-efficiency mode also includes community initiatives such as pop-up repair shops and clothing exchanges as well as local government initiatives such as 'free garbage days' in which households leave unwanted items on the curbside for other residents to glean.

According to the authors, the fourth mode of governance, energy-from-waste, focuses on taking advantage of any and all social, economic and environmental positive dividends that waste might afford. Examples include using waste as a way of organizing communities around shared values and commitments, a theme I return to in Chapter 6. Thus, waste-as-resource is as much about 'providing opportunities for skills development and employment and an entity to be governed through the mobilization of communities' (ibid: 2749) as it is about economically benefitting from waste-to-energy facilities that transform MSW into electricity and/or heating. In practice, though, waste-as-resource largely depends on linear economy EfW facilities and recycling facilities, and both are largely governed by private industry and require significant monetary investments from communities. These technologies and their management are also squarely constructed and managed for profit. As such, all four of Bulkeley et al's (2007) modes of governance are still entrenched in the neoliberal capitalist framework (with regard to individual responsibility and overarching focus on economic gains).

Given the force and enthusiasm with which local and federal governments, communities and non-governmental organizations around the world have embraced, championed and *become dependent on* recycling as the method of decreasing waste and improving the environment, it is important to examine what recycling actually entails and achieves. That is, it is vital that we study the research on the material benefits and costs of recycling compared with other waste management methods, as well as the social, economic and political consequences of recycling. Only by considering recycling's environmental footprint as well as how recycling operates within our cultural, economic and political systems are we able to meaningfully get beyond recycling rhetoric that has, since the 1970s, loudly declared that recycling is an environmental good.

Landfilling

Before we evaluate the benefits and costs of recycling waste, let's look at landfilling and incineration. While countries (especially in the globalized north) attempt to move away from these technologies, they remain vital to the global management of waste. Landfills, like all other forms of waste management (except refusal, reduction, reuse and refurbishment), have their own issues. The landfills of contemporary industrial societies include various amounts and kinds of seven million or so known chemicals (and the thousand new chemicals that enter into use each year), along with a full spectrum of organic matter, which includes the 14,000 food additives and the contaminants found in our food scraps. The liquid material, called leachate, into which organic landfill dissolves frequently consists of a heterogeneous mix of heavy metals, endocrine-disrupting chemicals, phthalates, herbicides, pesticides and various gases, including methane, carbon dioxide, carbon monoxide, hydrogen, oxygen, nitrogen and hydrogen sulphide (Hird, 2021). So, all of those diapers, food scraps, metals, holiday wrapping paper, polystyrene packaging, pieces of wood, liquids, refrigerators, pets (as well as their faeces and litter), batteries, chairs, pizza boxes, small appliances, fabrics and so on are re-stratified and compacted with other, rather less expected materials such as products of common industrial processes, like coal fly ash (of which over 50 per cent ends up landfilled; see Chertow, 2009), plastics (more than 308 million tons of plastics are consumed worldwide each year, most of which still end up landfilled; see PlasticsEurope, 2018), and food waste (over 97 per cent of which is landfilled in the United States; see Levis et al, 2010). To give one example, when food is recalled from Canadian supermarket shelves, it often ends up in landfills. XL Foods, Canada's largest food processor, processes over 40 per cent of the country's cattle and accounts for 30 per cent of the beef on store shelves. In the fall of 2012, approximately 5.5 million kilograms of beef presumed to be contaminated with *E. coli* was recalled, equivalent to 12,000 cattle. Of that, 500,000 kilograms were landfilled. When XL Foods wanted to reopen their plant to resume production, they were required to do a pilot test to ensure their corrective measures after the recall were effective. This test required the slaughter of 5,000 cattle, the carcasses of which were also landfilled after being tested for contamination, regardless of whether they had themselves been contaminated (Lougheed et al, 2016; Lougheed, 2017).

Landfills mix hazardous and non-hazardous waste. Over time, this waste may become unstable, or as scientists put it, 'variations [in leachate] may be cyclical, directional, stochastic, or chaotic' (Collins et al, 2000). Aerobic bacteria metabolize during the early life of a landfill, which produces material that is highly acidic and toxic to surface water. Anaerobic bacteria do the bulk of the metabolizing work deeper in the landfill's strata, producing leachate. And this leachate may travel vertically and horizontally within landfills and may continue to travel when it leaks out of the landfill. That is, leachate may percolate into soil and groundwater, where it moves into and through plants, trees, animals, fungi, insects and the atmosphere. It is this leachate that is also responsible for the methane that emanates from landfills. In 2010, landfills around the world produced nearly 882 million tons of CO_2 emissions, which amounts to about 11 per cent of all human-generated methane (Gies, 2016).

The task of the modern engineered landfill is to keep its contents in place, as static as possible, and for as long as possible. Myriad factors affect the efficacy of engineered landfills: the three primary subsystems of barrier, landfill operations and cover require considerable attention to siting, design, construction, operations and post-closure care (that is, the need to maintain a closed landfill for decades to centuries). Moreover, landfill regulations do not always address issues such as 'contaminants of emerging concern' (Celik et al, 2009; LaPensee et al, 2009; Rowe, 2012; Takai et al, 2000). Contaminants of emerging concern include chemicals such as bisphenol A (BPA) that has been used in many plastic products and is believed to mimic human oestrogen at low concentrations (LaPense et al, 2009; Takai et al, 2000), and polybrominated diphenyl ether (BPDE), which is an additive flame retardant in plastics, foams and fabrics that may cause liver, thyroid and neurodevelopmental toxicity – as well as new materials such as nanoparticles, which were not part of the waste stream at the time many landfill regulations were developed (such as US Subtitle D) (Islam and Rowe, 2009; Rowe, 2012; LaPense et al, 2009; Takai et al, 2000). In addition, calculating the environmental costs must include the climate change implications of using fossil fuels to transport waste to landfills (Rowe, 2012).

High-profile cases such as Love Canal in Niagara Falls, New York, speak to the catastrophic consequences of open dumping or badly engineered landfills. Love Canal was the subject of a 21-year Superfund clean-up of toxic chemicals dumped in a landfill. This toxic waste was covered over, and through a series of property sales ended up as the site of a residential community, including schools, playgrounds and so on (see Smith, 1982). The surrounding land and water were badly contaminated, and children and women were particularly affected by the contamination, with increased levels of particular cancers, neurological problems and birth defects (ibid).

Incineration

While, globally, open dumping and landfilling waste remain the most used forms of waste management, incineration is also deployed throughout the world. Traditional incinerators burn waste at very high temperatures. Incinerator gases entering the atmosphere via incinerator stacks may produce particulate pollution associated with cardiovascular and cerebrovascular mortality. Many of these emissions occur during day-to-day facility operations as well as the 'very high releases of dioxin that arise during start-up and shutdown of incinerators' (Thompson and Anthony, 2008: 2). Moreover, all incinerators produce waste ash that is potentially more hazardous than the waste that feeds the incinerator (Rowe, 2012). This waste ash is currently landfilled, which means concentrated levels of heavy metals are buried in the ground amid other landfilled materials as well as other constituents (such as calcium) that accelerate leachate collection system clogging, increasing the risk of leachate leakage (Rowe, 2012). When ash hydrates within landfills, it generates substantial heat that could compromise liner systems that could be, but typically are not, designed to accommodate these temperatures. In addition, to function properly, incinerators require a constant input of waste. For this reason, the phrase 'build the beast, feed the beast' is used to describe incinerators and leads to countries such as Sweden and Denmark importing waste from other countries (increasing carbon emissions via the transportation of this waste), and some municipalities to encourage their citizens to divert *less* waste from disposal in order to have sufficient waste for the incinerator. As one incinerator operator said, responding to China's plastic recycling import ban (see Chapter 4), "if we do the plastic ban, we would have to look for other more distant waste" (in Bahers, 2021: 10: translated from French). And we must keep in mind that waste-from-energy technologies are linear, not circular: once the electricity is used and/or the heating is used, it is gone.

Recycling

Given the clear limitations of landfills and incinerators, and wanting to move up the waste hierarchy, local governments have turned to recycling. Of all the materials that are diverted from landfills, food waste is unquestionably the most environmentally beneficial to recycle. When people compost food waste, different kinds of bacteria are transforming unwanted food into soil, which can then be used to grow more food. The more local the composting, the better for the environment, since transporting food waste to composting facilities is usually done with trucks that use non-renewable fossil fuels and are thus contributing to climate change. Diverting food waste from landfills is particularly beneficial because food waste is especially attractive to various kinds of aerobic bacteria that transform this waste into leachate that, combined with the various hazardous materials outlined above, and leaked into the environment, can cause serious human health and environmental harm.

Recycling's profit margin

However, once we consider recycling other materials, the situation rapidly complexifies. The first consideration is an outcome of the global neoliberal capitalist system that controls the extraction, manufacturing, circulation, retail, consumption and post-consumption of materials. This includes recycling: the materials that we put into our recycling bins – plastics, metals, paper products and so on – are *only actually* recycled if the recycling company makes a profit. Thus, one of the 'dirty little secrets' of recycling is that, due to constantly varying markets, materials the public believes are destined for diversion are in fact landfilled or incinerated: waste intended for recycling is often disposed of when recycling costs outweigh the profit derived from the recycled materials. Some recycling companies have the capacity to stockpile some materials while they wait for a more favourable market, but most often recycling companies simply move the materials into the (landfill or incineration and then landfill for the incinerator ash) disposal stream. Moreover, few regulations require recycling companies to declare what proportion of the materials they collect are actually recycled, and local governments are typically not legally required to monitor recycling companies – that is, once recycling companies collect materials, municipal governments are not required to know where the recycling companies move the materials, including whether the materials are taken to a recycling facility or to a landfill or incinerator.

The negative environmental impacts of using recycled materials

This leads to another important environmental consideration: the physical process of recycling materials. Recycling materials consumes a great deal of energy and mainly entails using non-renewable fossil fuels that pollute the soil and atmosphere and contribute to global warming (Center for Sustainability, 2012; MacBride, 2012). To illustrate, let's consider the results of an important meta-analysis study that synthesized the results of available research conducted on the life cycle of various packaging and food service ware materials. Jorge Vendries and his colleagues (2020) were interested in determining whether particular material attributes such as recycled content and recyclability produce lower net environmental impacts across the full life cycle of packaging and food service ware. Analysing the results of 71 studies that included over 5,000 comparisons of 13 impact categories, the authors' findings were illustrative of the complexities of recycling. The impact categories included: human toxicity, global warming, fossil energy, ecotoxicity, eutrophication, smog, acidification, PM formation, ozone depletion, mineral depletion, water consumption, land occupation and ionizing radiation.

The best-case scenario occurred with the same material that contained recycled content. In this case, just 20 per cent showed significantly lower impacts across all categories, while 74 per cent demonstrated marginally lower impacts (ibid: 5359). That is, when only one material is chosen, it has only a 20 per cent lower environmental impact when more of that material comes from recycled content than from non-recycled (that is, new or virgin) content. As Figure 3.2 shows, the meta-analysis significantly revealed that

Figure 3.2: Environmental impacts of recycled materials

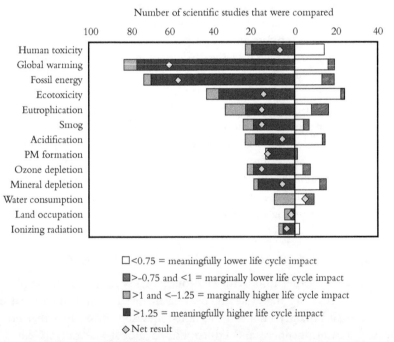

Source: Adapted with permission from Vendries, J., Sauer, B., Hawkins, T., Allaway, D., Canepa, P., Mistry, M. (2020) 'The significance of environmental attributes as indicators of the life cycle environmental impacts of packaging and food service ware', *Environmental Science and Technology*, 54: 5356–6. Copyright 2020. American Chemical Society.

the material matters more than recycled content in terms of environmental impacts: '[I]n fact, of 534 comparisons of packaging made from different materials and with different levels of recycled content, the packaging with higher recycled content had significantly higher impacts in 60% of the comparisons and significantly lower impacts in 21%'.

The results for recyclable packaging content using different material, biobased packaging using either the same material or different materials, compostable versus non-compostable packaging, and composting packaging versus other end-of-life options demonstrate a net significant *cost to* (that is, negative impact on) *the environment*.

Recycling creates waste

One of the problems here – in the fine print of recycling – is that recycling may release hazardous wastes into the environment through by-product emissions and/or require the use of toxic materials. Recycling paper, for instance, requires the significant use of toxic chemicals to remove ink and generates its own waste – sludge – that is more difficult to dispose of than paper (US Department of Energy, 2006). Using paper products requires harvesting wood, which in turn requires:

> manufactured heavy equipment and vast quantities of petroleum products. Composite wood products contain glues made from petroleum ... when wood is made into paper, only part of the wood, the cellulose, is used. The other part, the lignin, becomes waste. ... De-inking and re-pulping include a mechanical treatment. This damages the cellulose fibers. They break and the average fiber length decreases. Fiber length has a significant effect on the quality of the paper, therefore the fiber cannot be used indefinitely. (Baarschers, 1996: 181, 190)

Indeed, removing ink from the paper in the recycling process inevitably causes a loss of between 10 and 20 per cent of the fibre (ibid). In other words, each time paper is recycled, it produces a degraded product, which means that it cannot be recycled indefinitely.

A lot of materials cannot be recycled

A lot of the plastics waste that we put in the recycling stream *cannot* materially be recycled. Many types of non-durable plastics, such as those commonly used in food packaging, may be put in recycling bins by consumers, but they are diverted back into the disposal stream when they are transported to a recycling sorting centre. Yet another problem is that recycled materials are generally only suitable for one or two subsequent uses and usually only in lesser quality products. And without an ongoing market for many recycled materials, the 'consequence is that materials thought by the public to be headed for recycling end up in landfills' (Rowe, 2012: 6). For this reason, PlasticsEurope (2018) estimates that the vast bulk of recycled plastics ends up landfilled.

Moreover, in the case of plastics, the mechanical process of recycling plastics requires virgin resin. Mechanical plastic-to-plastic recycling involves 'sorting, washing, shredding/grinding, melting and pelletizing plastics waste into secondary raw material' (Greenpeace, 2020: 13). Not only is this most popular form of plastics recycling unable to repair recycled plastic degradation, leading to a reduction in recycled plastics quality and therefore reusability, but this has led to the mixing of virgin resin. As Chapter 4 details, plastic virgin resin means oil and gas. What we see, then, is an increase in oil and gas extraction with mechanical plastics recycling.

Recycling does not decrease extraction

So, another problem with recycling is that it has not led to a decrease in extraction. The case of electronics provides a stark example. Even if we were somehow able to recycle all of the metals and minerals contained in electronics such as laptops, smart phones and cars, the current demand for electronics has already outpaced the supply that recycled electronic waste could produce. As Figure 3.3 shows, as electronics demand increases, the minimum gap to be filled by primary resource extraction simply increases as well.

Figure 3.3: The growing gap between available metals and minerals and current and growing demand

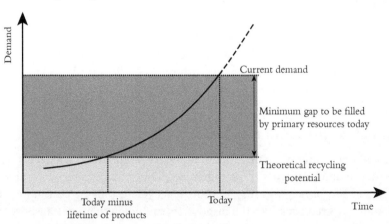

Source: L. Tercero Espinoza (2012) FP7 POUNARES Project POLINARES Working Paper, 20, Fraunhofer ISI

Moreover, evidence suggests that people actually produce more waste when they have access to recycling (Harris, 2015). Take a look at this graph developed by Scott Lougheed, one of my former doctoral students, and an expert on food recalls and waste (see Lougheed, 2017). Scott assembled data on consumption rates and the availability of recycling services in communities across Ontario, Canada's most populated province. As Figure 3.4 shows, there is a positive correlation between consumption and recycling, suggesting that in communities with recycling services, people may increase their consumption because they assume that the products and packaging they buy are being recycled.

The environmental costs of transporting recycling

In addition, the environmental benefits of recycling are particularly suspect when we consider that most recycling needs to be transported from a community to at least one recycling centre. That is, transporting waste to reach recycling processing facilities significantly mitigates claims that recycling is always the environmentally better option: 'Recycling requires collection, transportation and, at the recycling plant, industrial-scale processing. Transportation and processing mean energy consumption as we process wastes. The recycling industry is an industry and it inevitably produces its own industrial waste' (Baarschers, 1996: 187).

Materials often need to travel to more than one reprocessing centres in order to be recycled, which increases the environmental costs of using non-renewable fossil fuels, carbon emissions, road creation and maintenance issues (including the increased use of materials and pollution), increased production of waste transportation vehicles, and the increased risk of road accidents. As William Baarschers notes, 'after the diesel fuel is burned, the carbon dioxide and nitric oxides produced, we have waste tires, waste batteries, and waste trucks at the end of the line' (1996: 190).

To illustrate, Cassandra Kuyvenhoven and I conducted a study of transporting recycling in the community of Kingston, Ontario, where I work and Cassandra studied. We followed the recycling 'trail' for polystyrene (known as Styrofoam). Polystyrene is the most difficult material for the Kingston Area Recycling Centre (KARC) to sell, mainly because of its high volume and low weight (Schliesmann, 2012). KARC used to contract with an Asset Recovery plant in North Bay, Ontario (approximately 454 kilometres away from Kingston), which received Kingston's approximately 47 annual tons of polystyrene (ibid). The plant in North Bay super-compressed the polystyrene and shipped 60 per cent of this compressed polystyrene to South Korea and the other 40 per cent to the United States (ibid). The super-compressed polystyrene was then refashioned into household baseboard products and picture frames, and then

Figure 3.4: Ontario municipalities waste diversion x per capita waste generation

Per capita waste generation (kilograms)

Diversion rate

◆ Diversion rate

— Linear (diversion rate)

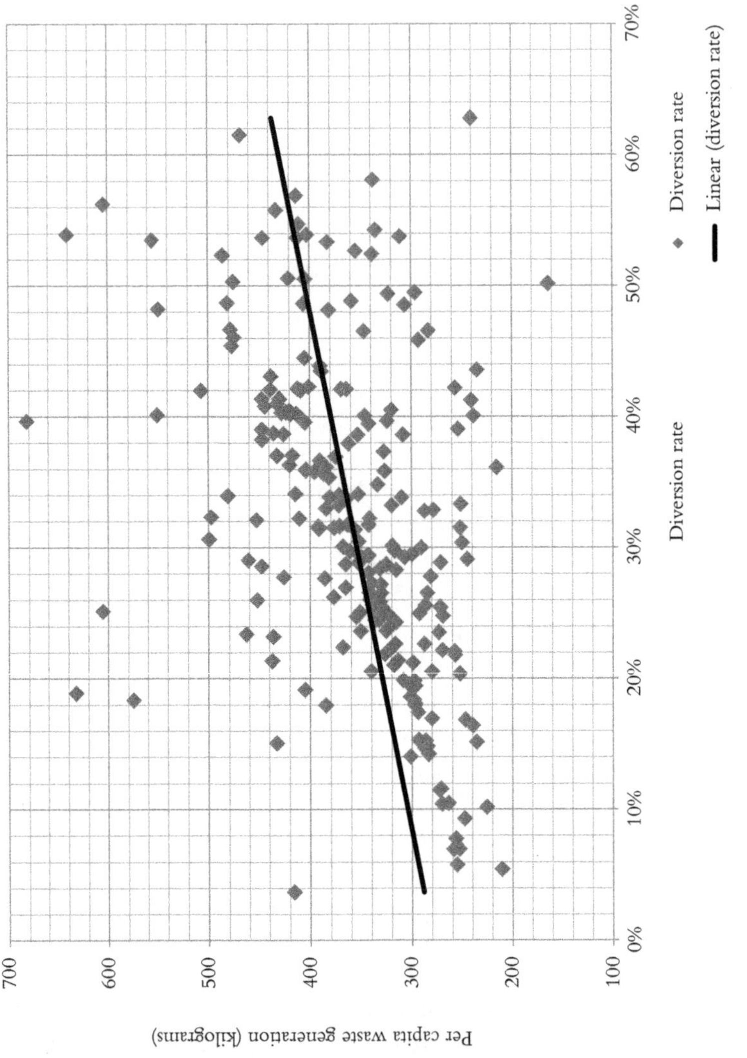

Source: Scott Lougheed.

shipped back for sale in North America (and presumably then eventually discarded). When we factor in the environmental costs of transportation, it is difficult to argue that the environmental costs of transporting polystyrene across the globe (to be eventually landfilled or openly dumped) are lower than the environmental costs of landfilling the polystyrene locally (or better yet, refusing to allow companies to use polystyrene, refusing to buy products that come with polystyrene packaging or refusing to buy certain products altogether).

Indeed, the transportation routes of recyclables can vary widely. As of 2012, the closest film plastics (that is, shrink wrap and shopping bags) recycler to Kingston was Waste Logix in Brechin, Ontario (approximately 300 kilometres away). However, after the waste was shipped to Brechin, Waste Logix then shipped the plastics to a Texas company (approximately 2,700 kilometres away) that processes, sorts and pelletizes the plastics (Schliesmann, 2012). On my most recent visit to KARC in October 2019, an employee told me that the polystyrene is now being transported to Indianapolis, Indiana, for chemical reprocessing, which is approximately 626 kilometres further away than the recycling plant in North Bay, Ontario (and approximately 1,080 kilometres away from Kingston).

Unless a community has its own in-house recycling facility, it will need to ship 100 per cent of its recyclables to another region. According to a greenhouse gas calculator, based on Statistics Canada trucking data, a heavy truck emits 114 grammes of carbon dioxide equivalents per tonne-kilometre (114g CO2e/t-km) (Government of Canada 2016). A typical waste collection truck will "have a curb weight of 13 tonnes and a 9.5 ton payload capacity (the carrying capacity of a vehicle, usually measured in terms of weight), but gross vehicle weights vary up to 27 tonnes" (OWMA, 2016: 10). These curbside pickup trucks are the most energy intensive vehicles on the road, due in part to their weight, low speed, stop-and-go cycles, and high idling times (ibid). According to Environment Canada, MSW collection vehicles are classified as heavy-duty diesel vehicles (HDDV), which is the same classification assigned to long-haul tractors (Agar et al, 2012). The most significant emissions are carbon monoxide (CO) and carbon dioxide (CO_2). The rates are also given for nitrous oxide (N_2O), methane (CH_4), nitrogen oxides (NO_x), and particulate matter (PM10) greater than 10 micrograms (μg) in diameter. These greenhouse gas emissions (GHGs) are significant contributors to climate change (Government of Canada, 2016). Thus, when Kingston's waste is directly disposed of in landfills, it remains in North America – primarily in Ontario (OWMA, 2016). When Kingston recycles its waste, the distance is considerably greater – thousands versus hundreds of kilometres.

In other words, diversion in the form of recycling that relies on mid- and long-haul transportation contributes to the problem of global warming.

According to Naya Olmer et al's (2017) analysis of container fleet CO_2 emissions, container ships alone contribute about 23 per cent of human-produced carbon emissions (more than oil tankers, which account for 13 per cent). In 2015, international container ships produced approximately 31,419 million tons of CO_2.

Waste transportation involves several environmental and health-related issues, including climate change due to greenhouse gas emissions, a reliance on non-renewable fossil fuels and a substantial expenditure of energy, and various safety risks incurred through contamination, spills, leaks and injuries (Thomson, 2009; Eisted et al, 2009; Gregson et al, 2010). The further or more complex the transportation route, the higher the probability that there will be negative impacts on human and environmental health and safety (Thomson, 2009; Eisted et al, 2009; Gregson et al, 2007). Garbage trucks account for 5.12 human deaths per 100 million miles travelled (Rosenfeld, 2013), compared with other light motor vehicles (cars and light trucks), which account for 1.13 human deaths per 100 million miles travelled (Insurance Institute for Highway Safety, 2016).

Furthermore, waste collectors are routinely exposed to hazardous conditions, including coming into contact with toxic materials, inhaling pathogens and dust, and handling sharp objects (Yong Jeong, 2016). Although deaths in the occupation are relatively rare, 'ergonomic injuries, such as back strain, are commonplace and cuts from sharp objects and exposure to bacteria and toxins are always a threat' (Tibbetts, 2013: 185). According to Janice Tibbetts, 'if you're on the job for five years ... I would say it's a pretty safe statement that you're going to get some kind of injury' (ibid). According to the Canadian Union of Public Employees, garbage collection is one of the most hazardous jobs, with injuries sustained by approximately 35 per cent of garbage collectors in Canada each year (ibid).

The low-value limit

All of this collecting, sorting, transporting and mechanical and chemical re-processing takes place to produce a fairly low-value product that will be retailed, consumed and then discarded. As Jodie Morgan observes: '[P]lastic in general is a fairly low-value product. And then we spend a lot of money collecting that product, sorting that product, processing that product all so it can go back into ... a relatively low-value end product' (in Chung, 2019: np). According to a recent life-cycle analysis study of the environmental footprint of polystyrene, researchers Brooke Marten and Andrea Hicks (2018) note that just the re-expanding and shaping stages of polystyrene recycling takes 30 per cent of the energy used in the life cycle of this material. They note that we need to consider the fine print of claiming that any recycling process is better for the environment: recycling technologies use energy and chemical solvents, both of which have an environmental footprint. As Hicks notes: 'There's that whole reduce, reuse, recycle. And I think what people really miss is, it's actually a hierarchy. We're supposed to reduce what we're using, then reuse, and the last case is recycling' (in Chung, 2019: np).

Externalizing recycling costs to consumers

There are financial, social and political costs as well. One of the main problems with recycling is that it externalizes the monetary cost of waste on to consumers. As Max Liboiron observes:

> The Container Corporation of America sponsored the creating of the recycling symbol for the first Earth Day in 1970 (Rogers, 2006, 171). The American Chemistry Council, the world's largest plastics lobby, enthusiastically testified in favour of expanding New York City's curbside recycling program to accept rigid plastics (ACC 2010). Recycling is a far greater benefit to industry than to the environment. ... Industry champions recycling because if a company has reusable bottles, for example, it has to pay for those bottles to return, but if it makes cheap disposables, municipalities pick up the bill for running them to the landfill or recycling station. (2013, 10–11)

In other words, recycling deflects attention away from Extended Producer Responsibility (EPR) regulations and policies, which require product and packaging producers to take financial and material responsibility for all of this stuff (for more on EPR, see Chapter 6). Put succinctly, recycling works very well for the extractive and manufacturing industries and for retailers, and it works very badly for consumers, citizens and the environment. As such, recycling reveals a significant imbalance of power between extractive, manufacturing and retail industries and the public, who are shouldering the financial burden of industry's waste production.

Municipalities around the world have largely acceded to industries' externalization of financial, behavioural and moral responsibility, and often spend significant sums of money on recycling education programmes targeting their constituents. For instance, the City of Kingston where I work 'rel[ies] heavily on our ability to educate our residents and encourage them to change their usage patterns and behaviours' (2013: 6). These education initiatives may well be partially or fully financed through taxes, which means that consumers are paying to take more responsibility for waste they have not produced. Sometimes manufacturers will partially or fully foot the bill for these education drives, such as when Stewardship Ontario – a consortium of industries that produce the packaging and products that go into the province's recycling bins – provides funding to schools to encourage children to increase their recycling behaviours. In addition, these education initiatives almost exclusively adopt a deficit model (Roth and Désautels, 2004), whereby the assumption is that recycling rates will increase with increased education (for a critique of the Deficit Model, see Hird, 2011). These educational initiatives amount to a fraction of the cost that manufacturers would pay if they were regulated to take financial responsibility for the waste they currently foist on to consumers, so bearing the cost of 'educating the public' – often with much advertising and fanfare – is a much smaller financial burden than actually paying to manage the waste these companies produce.

The moral economy of recycling as a public problem

And so, when it comes to recycling, the devil, it may be said, is in the details. According to Annie Leonard's *The Story of Stuff* (2011), 99 per cent of purchased consumer goods in the United States end up being thrown away within six months of purchase. This means that whatever waste is recycled will almost certainly – *and typically after only one recycle* – turn into waste that will be transported to a landfill, an incinerator or EfW facility. There are very significant environmental costs, then, of recycling a material that will likely only get one more use before disposal (recall the low-value limit).

Recycling does not only cost individuals, families and households in monetary terms. We also pay for recycling as part of the 'moral economy' of our societies. The term was coined by E.P. Thompson in his 1963 book, *The Making of the English Working Class*. Thompson's aim was to understand why poor people in feudal England rioted against the shift towards capitalism, which included dramatic changes in laws and practices. For example, feudalism allowed for common lands to be used by peasants to grow food for their families. Capitalism introduced an 'enclosure system' whereby wealthy individuals and families claimed ownership of these common or 'waste lands', thereby depriving poor people of a vital means of survival (Clark and Clark, 2002). Capitalism's emphasis on the (so-called) free market and laissez-faire governance ensured that wealthy individuals and families increasingly profited from the poor's vulnerabilities. For Thompson, the ensuing riots that spanned the 1530s to 1640s were a reflection not only of these dramatic economic changes but were undergirded by strong social norms concerned with the mutual obligations and responsibilities between the propertied wealthy class and the peasants who worked their lard. That is, the protests were born of a moral outrage that the feudal system – which, however inequitable it was, had organized generations of poor and rich within a system that afforded at least some protections to poor families – was being replaced by a system of naked and unbridled profit (Stehr et al, 2009).

Thus, we may think about the relationships between various sectors of society, from citizens and consumers to local government, retailers and manufacturers as not only engaging in what we traditionally think of as the regulations, policies and practices through which monetary goods and service transactions take place, but also that these relationships are part of a moral economy as well. And when it comes to waste, much of this moral economy centres on individuals shouldering the responsibility for disposal and recycling. Local communities everywhere have initiated various 'good citizen' inducements to recycle through programmes that publicize individuals and households that recycle at higher rates. For instance, in Kingston where I work, the 'Remarkable Recyclers' campaign (John, 2012: 1) recognizes households recycling at least 75 per cent of their waste with a special badge on their curbside bins (City Hall Public Notice, 2013). Kingston also offers initiatives like 'Train the Trainer' that provides educational training to university student representatives to 'help them provide accurate information to tenants when incorrect materials are seen being placed at the curb' (Kingston City Council, 2013: 37). The city has also introduced various educational programmes and events, such as 'Pitch-In Kingston', which is a community clean-up programme whereby individuals volunteer their time to pick up litter in their community.

Scott Lougheed and I conducted a study in Kingston analysing the local media's coverage of waste management issues. Of a total of 81 newspaper reports we sampled between the years 2008 and 2013, most (59, or nearly 73 per cent) of the newspapers' reports and opinion pieces (where members of the public write to the newspaper to express concerns) focused on waste as a matter of individual responsibility. Occasionally, the newspaper articles took a more in-depth approach that attempted to move the analysis beyond recycling. The primary example of this was an article written in the *Kingston Whig-Standard*. Ostensibly about the origins of recycling in Ontario, the report went on to explicitly identify recycling as a profit-making business that 'reinforces the notion that citizens, as consumers and recyclers, are crucial to the manufacturing chain' (Schliesmann, 2011b: 4). In a fascinating follow-up article, the same columnist detailed experts' concerns with recycling. The article drew particular attention to how industry tied recycling to commodity markets from the outset, establishing monetary incentives for municipalities to prioritize recycling rather than waste reduction into the system (Schliesmann, 2012: 4). As Thomas Naylor bluntly put it, 'the recycling game is a con' insofar as it sustains the consumer-based economy and does not legislate manufacturers to either reduce/eliminate packaging, or take back goods when the consumer no longer wants them (quoted in Schliesmann, 2011a).

With these infrequent exceptions, most media reports and citizen editorials fit squarely within a moral economy that emphasizes individual responsibility, with an emphasis on self-surveillance and the surveillance of others as 'good citizens'. For instance, one Kingston resident wrote that 'having a little enviroguilt [sic] can be a good thing' (Switzer, 2008: 1), and another writes of a need to 'curb her appetite for plastic' (Browne, 2008: 1). Several opinion editorials were devoted to Kingston residents monitoring each other: neighbours and fellow residents became 'bad citizens', and authors offered advice about how citizens could transform themselves from 'bad' to 'good'. For instance, one resident advised residents to take coffee cups from cafes home and recycle them, as she did (Toomey, 2008: 1); a resident of a neighbouring township defined people who do not want to use clear bags for recycling as 'anti-recycling' and wanting to 'hide (stuff)' and as people who were 'not true recyclers' (Jefferson in Edmiston, 2010: 1).

Thus, there is, as Kathryn Wheeler and Miriam Glucksmann point out, a strong moral economy that operates such that the 'responsible "citizen-consumer" is motivated to act because of his or her commitment to moral/political projects rather than in line with his or her selfish desires' (2015: 143–4). Many countries have adopted recycling as a central part of their moral economy and emblematic of the 'good environmental citizen': for instance, studies show that countries such as the United Kingdom and Sweden have adopted this discourse wholesale (ibid; Skill, 2008). Indeed, Skill (2008) demonstrates that around the globe, 'recycling is the most common action that households regularly performed' (in Wheeler and Glucksmann, 2015: 153). In the United Kingdom, this moral economy extends to saving public money through recycling. And in Sweden, Swedish-raised respondents distinguished themselves from immigrants as the 'irresponsible other', based on the perceived lack of proper recycling performed by immigrants (Sayer, 2000). Around the globe, then, ordinary people are shouldering the financial and moral burden of a system that was rigged by manufacturing industries from the start.

4

The Public Problem of Plastics

Introduction

When I was about nine years old, my mother patiently taught me how to knit. As we sat in her tiny sewing room and she showed me the impressive collection of yarn, knitting needles and patterns that she had collected throughout her life, my mother handed me a rigid spherical ball – two equal parts that could be screwed together, hollow on the inside and with a small circular hole in one of the halves. My mother explained that a plastics company had invented this ball ostensibly to keep the wool from either unravelling or being soiled. She said that after World War II companies were inventing anything and everything to make with plastics. Even as a child of the later 20th century, immersed from birth in a plastics world, I remember thinking that this was a strange object. Many years later, as a waste studies researcher, I find this object in equal measure both unsurprising and shocking. Plastics characterize our contemporary society. They proliferate within our homes and workspaces, are embedded in the fabric of our clothing, footwear, personal hygiene products, our bicycles, cars and other modes of transportation, our food, and even the human placental barrier. Like the proverbial lobsters in a slowly heated pot, we have become so accustomed to plastics' incessant overabundance that we are overwhelmed by our current plastics waste crisis.

From the giddy 'miracle' years of plastics and their seemingly limitless applications and everyday benefits to the public's current disaffected dependence on plastics, this chapter focuses on how plastics have been framed by, primarily, the oil and gas industries that provide the raw materials needed to make plastics, and governments trying to appease publics' growing concerns with plastics' negative impacts on human health and the environment. Over 170 countries, from Canada to Kenya, the United Kingdom to China, Zimbabwe to India, have signed on to 'significantly reduce plastics' by the year 2030 (Masterson, 2020: np). At the same time, oil and gas industries have steadily (and sometimes exponentially) increased their plastics production, and make no secret of their plans to increase plastics production further. This chapter examines how the oil and gas industries are deploying a waste-as-resource frame to maintain their hegemony within the energy market, as the primary means by which they continue to produce plastics despite concurrent government pledges to reduce plastics. While governments and environmental organizations highlight China's plastics ban declaration in 2017 (implemented in 2018) and the Basel Convention on the Transboundary Export of Hazardous Wastes as key regulatory mechanisms through which to restrict plastics production and thereby decrease the global circulation of plastics, oil and gas industries are zealously framing plastics as a key resource: not only in securing a flourishing economic future (and *ipso facto* the security and sovereignty of nation-states) but also an environmentally sustainable one that will resolve our plastics problem through recycling.

This chapter illustrates the waste-as-resource frame with a case study of Canada's delay in ratifying the Basel Convention Amendments and its rather scrambled Bilateral Agreement with the United States that allows for the continued transboundary movement of plastics for the purpose of recycling. I particularly emphasize how both the oil and gas industries and governments attempt to restrict the parameters of public participation in the industries' continued dominance in energy policy and practice in what amounts to a sleight-of-hand manoeuvre to assure publics that the plastics problem is resolvable through continued plastics production.

The promise of plastics

The history of plastics is, in many ways, the iconic story of human scientific and technological progress: of humanity's triumph over nature and a crucial link in the chain leading to modernity. The first known use of polymers – natural rubber material – dates back to approximately 1600 BC (Hosler et al, 1999). In 1839, Goodyear invented vulcanized rubber for automobile tires and Eduard Simon invented polystyrene (Andrady and Neal, 2009). In 1856, Alexander Parkes created Parkesine. Leo Baekeland developed the first synthetic plastic in 1907, which he named Bakelite. Plastics development greatly expanded in the first half of the 20th century to include: polyethylene terephthalate used in clear water and soft drink bottles; high-density polyethylene used to make, for instance, laundry detergent and bleach containers; low-density polyethylene made for use as plastic shopping bags and cling film; polyvinyl chloride used in making products like toys (such as the iconic 'rubber' ducky); polypropylene made into products like margarine containers and some bottle caps; and polystyrene, which most people know as Styrofoam (PS) and is used, for instance, in product packaging.

As proponents of plastics such as PlasticsEurope and the British Plastics Federation like to remind us, the popularity of plastics owes much to its material versatility: plastics withstand a large range of temperatures, are lightweight, make excellent thermal and electrical insulation but are also able to conduct electricity, be combined and moulded into an almost infinite number of shapes, and are amenable to production in staggeringly high numbers. And, indeed, thanks to this sheer versatility – and also to their abundant feedstock – plastics have become ubiquitous in the energy, health care, building and construction, agriculture and transportation sectors, as well as in sports, electronics and packaging.

Plastics, and their feedstock oil, played a very significant role in capitalism's advancement. Exploring the transition from ivory to plastic hair combs, Susan Freinkel (2011) argues that the cheap price of plastics products vastly expanded their market. As Jeffrey Meikle observes about the rise of celluloid film, '[b]y replacing materials that were hard to find or expensive to process, celluloid democratized a host of goods for an expanding consumption-oriented middle class' (1995: 14). Plastics were integral to mass production and its sister, mass consumption. According to the 'New plastics economy: rethinking the future of plastics' (2016) report produced by the industry-friendly World Economic Forum, Ellen MacArthur Foundation and McKinsey & Company, we are producing more plastics than ever before: from 15 million tons in 1964 to 311 million tons in 2014, with an expected doubling in production over the next 20 years. Packaging alone accounts for about 26 per cent of plastics production.

Facing increasing pressure from environmental groups such as Greenpeace, Friends of the Earth, Plastic Pollution Coalition, Plastic Change and the World Wildlife Fund, governments and concerned members of the public, the contemporary plastics industry (that is, the fossil fuel and chemical industries) is hard at work making the argument that plastics are the lesser of environmental evils. During the heady years of post-war capitalist expansion during which oil and gas were confidently declared to be limitless resources (Leggett, 2005), plastics also promised to solve the problem of natural resource depletion (other-than oil and gas), such as in the case of the near extinction of elephants, whose tusks were increasingly harvested for popular items like billiard balls and piano keys (Freinkel, 2011). Indeed, as plastics production increased, oil and gas companies began touting plastics as significant contributors to environmental conservation. Today, the plastics industry repeatedly points to decreases in food waste due to advances in plastics packaging that extend food's shelf life. Substituting heavier materials in cars and airplanes with plastics saves fuel. Plastics are also advertised as a means of carbon-sinking, thus lowering our carbon footprint (Roush, 2019). Proponents of plastics, such as Amy Hodgetts of Omega Plastics (2018), point out that plastics used in construction, such as insulation and double-glazing, conserve heat and thus save energy. And according to the Bank of England

(2019), which is moving from paper to polymer banknotes, plastics money is more durable and therefore carries a lower carbon footprint.

To wit, the 2014 report by Trucost, supported and financed by the American Chemistry Council, argues that the environmental costs of plastics packaging and consumer goods is four times less than the environmental costs of plastics alternatives or some $139 billion to $533 billion annually (Lord, 2016). Commenting on the report, Steve Russell, vice president of plastics for the ACC, lauded that "[w]e now have a fuller picture of the environmental benefits of using plastics. From lighter, more fuel-efficient cars to smart packaging that helps our favorite food last longer, our industry is committed to ongoing innovations that will advance sustainability across major market sectors and the globe" (American Chemistry Council, 2016: np).

A similar study by Franklin Associates, also prepared for the ACC, and using theoretical modelling, found that replacing plastics packaging would increase 'energy use, water consumption and solid waste, as well as increase greenhouse gas emissions (GHG), acidification, eutrophication and ozone depletion' (2018: np). And each year, the Canadian Plastics Industry Association funds a Plasticurious video contest that encourages children between the ages of 14 and 18 to compete for a CAD$1,000 prize. As the association's vice-president of sustainability, Joe Hruska, tweeted in 2019, the contest provides an opportunity for teens to 'educate themselves and their peers about how plastics power their life' with this 'amazing 21st century material', adding for good measure that 'we're excited to learn more about the role of plastics for Canadian teens!' (in Tompkins, 2019: np).

While pro-plastics arguments continue to extol the versatility of plastics to substitute for precious natural resources (that, of course, assumes no reduction in products), from animal parts to minerals, contemporary plastics proponents have – under pressure – turned their attention to addressing public concerns about plastics waste. That is, as human health concerns about exposure to plastics continue, plastics pollution and waste have emerged as the single largest environmental concern after climate change. As Thomas Le Roux and others have pointed out, the more technological progress, the more waste: 'After 1945, the plastic boom completely changes the waste question. Today, plastic waste has become the main issue in the waste debate' (Ducros 2019: np; see also Le Roux, 2016). And the plastics pollution statistics are alarming: according to Roland Geyer and colleagues (2017), approximately 6,300 million tons of plastics were generated between 1950 and 2015, and a further 302 million tons of plastics waste were produced in 2015 alone. Of this, only 600 million tons has been recycled, and only 10 per cent of that recycled more than once. Industrial plastics have a maximum 35-year service life and plastics take from five to 1,000 years to decompose into microplastics.

And microplastics effectively last forever (O'Neill, 2019). Even according to the World Economic Forum, a truck load of plastics are dumped in the world's oceans every minute (Pennington, 2016), and in 2017 the United Nations declared ocean plastics to be a 'planetary crisis'.

Scientists are continuing to explore the impacts, typically in isolation rather than synthesized, of the chemical additives used to make plastics such as bisphenol-A and phthalates that are carcinogenic and endocrine-disrupting (Greenpeace, 2020). Plastics waste has permeated soil, oceans and the atmosphere. Most of the global production of plastics is devoted to plastics packaging. Packaging is also the single largest source of plastics waste in the environment, and packaging is most often designed to be single-use (Plastics Europe, 2018). And according to a 2019 Center for International Environmental Law report, by 2050 global greenhouse emissions from the plastics production and consumption life cycle will account for 10 to 13 per cent of our planet's remaining emissions 'budget'.

In addition to the climate impacts of plastics waste, there are climate impacts at every point of the plastics life cycle. The production process requires fossil fuel input to make the plastics but also to maintain the high temperatures necessary for refining and manufacturing (Royer et al, 2018). Methane, which is both a fuel and a potent greenhouse gas, tends to leach during drilling, transport and refining, making it a 'hidden' source of pollution because methane is not typically measured as a greenhouse gas emission within the oil and gas industry. Emerging research has also shown that polyethylene plastics release greenhouse gases when breaking down in oceans, potentially interfering with tiny algae plants that play a crucial role in helping oceans absorb excess carbon from the atmosphere (ibid). Even when recyclable plastics make it to (plastics) recycling bins, much of it ends up in landfills, while the remaining 12 per cent is incinerated to produce single-use energy, potentially leaking toxic fumes into nearby communities and exacerbating the volume of carbon pollution in the atmosphere. Indeed, the dramatic increase in plastics production has resulted in a 15 per cent increase in emissions from 2012 to 2018 (CIEL, 2019). And, as Rebecca Leber found:

In 2020 alone, the Centre for International Environmental Law, using Environmental Integrity Project data, estimated that plastics production contributed the equivalent of 189 large coal plants. And if plastics production continues apace, the sector is on track to reach the equivalent annual pollution of 295 large coal plants in the next 10 years, and double that by 2050. And a 2018 International Energy Agency report indicated that carbon pollution from the petrochemical sector *will increase by thirty percent by 2050 over the sector's current rate.* (2020: np, emphasis added)

The plastics tide turned

Countries, particularly in the globalized north, have long relied on exporting their waste. The European Union is the largest exporter of waste, followed by the United States as largest single-country exporter (BBC News, 2019). It is now a rather standard investigative news piece to feature journalists picking through colossal piles of garbage in Cambodia or some other developing country and finding waste from the United Kingdom, France, Germany and elsewhere. In just one such example, the Basel Action Network discovered that 40 per cent of the electronic waste exported as recycling ended up in developing countries as waste, with 93 per cent of this waste ending up in China (Basel Action Network, 2016). The export of waste and materials labelled as 'recycling' from the globalized north to south has steadily increased for decades. Before 2018, China imported the greatest percentage of plastics and other recycling (such as copper, aluminium and paper). In 2016 alone, the United States exported 1,500 shipping containers of plastics and other scrap (such as paper and metals) *per day* to China (Flower, 2016). While China had already attempted to limit scrap imports before, most notably through its Operation Green Fence in 2013, in 2017 China very publicly declared an import cessation – calling it the National Sword Policy – sending many countries into a panicked tailspin. Malaysia largely picked up the waste exports that would have gone to China, and like other countries that import waste such as Vietnam, South Korea, India, Taiwan, Indonesia and Thailand, it is now experiencing the same kinds of issues that China faced. Much of this ever-increasing recycling is actually contaminated and cannot be recycled: that is, much of the plastics labelled as recycling at the point of origin is actually dirty plastics that have been mixed with regular non-plastics garbage. And this means that the receiving countries cannot glean any profit from recycling and are burdened with increasing waste disposal problems. As Chapter 1 details, the international fiasco between Canada and the Philippines is the tip of a colossal, and ever-growing, waste export crisis. Many countries are scrambling to find ways of continuing this linear waste export chain as well as deal with an increasingly informed and discontented public demanding change.

The 'Basel Convention on the Control of Transboundary Movements of Hazardous Wastes and their Disposal' is a significant response to the waste export crisis. Adopted in March 1989 and ratified in May 1992, the convention is a supra-national agreement designed to eliminate the movement of hazardous waste between countries, and particularly from wealthy to poor countries.[1] It specifically requires written consent from importers. A total of 189 countries, as well as the European Union, signed on to the treaty in 1989, but the United States and Haiti have yet to ratify it. In 2018, Norway proposed adding plastics waste to the convention's Annex II in what has come to be known as the Basel Ban Amendment. To date, 97 countries have ratified this amendment that bans the export of waste from a list of (mostly OECD) countries to (mostly non-OECD) developing countries, and significantly, includes recycling. Effectively, the amendment is designed to stop the export flow of waste from rich to poor countries under the guise of 'recycling'. Some countries, notably Australia, the United States and Canada, objected to the amendment. One objection is that it could pit the Basel Convention Amendment against the World Trade Organization (WTO) in that a country could claim that its ability to trade (in this case plastics or other hazardous wastes) is restricted because it is on the 'wrong' side (that is, the exporting side of the OECD list).[2] Canada specifically objected on the grounds that, first, processing imported recycling materials provides valuable employment for people in developing countries and second, that Canada lacks the recycling infrastructure to recycle the waste that it produces. The first claim, that ending the export of waste to poorer countries would negatively affect people whose livelihoods depend on waste-picking is a tricky argument to make. Effectively, Catherine McKenna, Canada's minister of environment and climate change at the time, argued that it is acceptable to export waste from wealthy parts of Canada to poor countries where children, women and men work in conditions whose health and safety violations and standards would make the same work in Canada illegal.[3] After considerable backlash from the Canadian public, the Canadian government signed on to the Basel Convention Amendments, but with an interesting – and very concerning – twist, that Canada's second objection foreshadowed, and that is taken up in the next section of this chapter.

In response to public pressure, a number of countries have already, or are currently, introducing various plastics waste 'bans' for certain products. For instance, there has been an increasing domino-effect ban on disposable plastic bags: Bangladesh led the ban in 1998, followed by several Indian states in the late 1990s. Taiwan phased out plastic bags in 2002, and South Africa started charging for plastic bags in 2004. The cities of San Francisco, Seattle, Los Angeles and Portland in the United States, and Mexico City in Mexico all banned single-use plastic bags in 2007. Provinces and states such as Manitoba, Hawaii and North Carolina, and all states in Australia except New South Wales, as well as whole countries including Italy, China, Bangladesh, Rwanda, Kenya, the Congo and South Africa have banned single-use plastic bags. In 2018, the United Nations Environment Programme published a report called 'Single-use plastics: a roadmap for sustainability', and the EU proposed a ban on single-use plastics such as straws and cutlery. As Chapter 5 details, the COVID-19 pandemic has astronomically increased the use of single-use plastics products, especially PPE, and some countries are announcing delays to the implementation of single-use plastics bans as the COVID-19 pandemic continues.

While governments are publicizing that these single-use plastics bans are evidence of their commitment to reducing plastics waste, criticisms are quickly mounting. Greenpeace Canada (2020), for instance, observes that the food service ware products scheduled to be banned in Canada in 2021 account for less than 1 per cent of plastics demand. Moreover, as Chapter 3 details, substituting bio-based or compostable FSW is often worse when we take into account a fuller environmental costs ledger that includes things like global warming, mineral depletion, ecotoxicity and smog (Vendries et al, 2020). And it appears quite clear that the plastics reduction strategy that countries in the globalized north are heavily relying on is not spearheaded by single-use plastics bans, reduction or reuse initiatives, but rather by significant increases in plastics recycling.

Oil by any other name: plastics recycling and the Green Energy Plan

As we are increasingly bombarded with alarming statistics on the negative human health impacts of plastics and their contamination of the environment, one of the subtle and therefore insidious features of plastics is that the name masks the source: oil. Almost all (99.9 per cent) plastics are derived from oil and gas. And yet the rhetorical separation of plastics from oil has served to disassociate all of the criticisms of the environmental and human-health costs of oil from plastics. As such, the issue of plastics waste cannot be isolated from the oil and gas industries themselves and the major producers of carbon emissions responsible for climate change. And while there is some argument over whether the environmental costs of plastics should be kept distinct from that of climate change, Zhu (2021) argues that all of the plastics that have ever been produced – some seven gigatonnes – is part of the carbon cycle. Factoring in plastics to our pressing climate change issue makes sense because, as Zhu notes:

> Plastic and climate are two sides of the same coin: the majority of plastic polymers are made from petrochemical feed-stocks and their raw materials for synthesis are ethylene and propylene. These compounds are derived from naphtha, one of several chemicals refined from petroleum. What else is refined from petroleum? Gasoline, the fossil fuel we burn for energy that emits greenhouse gases. (Ibid: np)

While these are different polymers, their feedstock is the same: petroleum.

According to some news media, the oil and gas industry is in crisis. In 2020, crude prices plummeted as the novel coronavirus swept the globe – combining a global humanitarian crisis with a fossil fuel supply shock and an unprecedented drop in demand (Barbosa et al, 2020). At the same time, accelerated societal pressure and a resurgence of international commitments to the Paris Agreement have seen an annual increasing share of renewables, such as wind and solar (IEA, 2020a; Reuters, 2021). In addition, more than 17 countries have set targets to phase out internal combustion engine vehicles (IEA, 2020b). As a result, oil giants have been forced to sell billions of dollars in assets to pay off debts from the oil stock price crash (Bousso, 2020), and newly elected US President Biden has issued an executive order that effectively terminates the Keystone XL pipeline project between the United States and Canada (Mabee, 2021).

Given the significant and by now well-documented environmental costs of oil and gas (and plastics production), coupled with the international move towards green energy and plastics reduction through single-use plastics bans and other measures, we might expect to see industries and countries declaring 'just transitions' from oil and gas (and their offspring, plastics) towards sustainable energy production and overall plastics reduction. And as the previous section details, companies and corporations are certainly making declarative environmental pledges. General Motors has declared its 'Path to an All-Electric Future' (Abuelsamid, 2021: np). This car company has pledged to only produce electric vehicles by 2035 and for their car manufacturing to be carbon neutral by 2040 (ibid: np). Proctor & Gamble announced goals to reduce virgin plastic usage by 50 per cent and reach 100 per cent recyclability or reusability by 2030 (waste360, 2021: np). And as I write this, the COP26 is coming to an uncertain conclusion, with various promises made to move towards 'clean energy'.

And yet, petrochemicals – chemicals produced through petroleum processing and that in turn produce plastics – promise to bolster the profits of extractive companies for decades to come. Indeed, the International Energy Agency (IEA) projects that plastics resins are the only area of fossil fuel demand that will *increase* in the coming years. As Fatih Birol, executive director of the IEA, admonishes: 'Our economies are heavily dependent on petrochemicals, but the sector receives far less attention than it deserves. Petrochemicals are one of the key blind spots in the global energy debate, especially given the influence they will exert on future energy trends' (IEA, 2018: np).

According to Michael Dent, '[o]wing to the push from industry, global revenues from plastics recycling – and chemical recycling, especially – is set to expand by about 30% annually over the coming decade. Plastics recycling will grow from $48 billion in revenues today to $162 billion by 2030' (2020: np). And indeed, the demand for plastics – the most familiar of the petrochemical products – has outpaced that of all other bulk materials (such as steel, aluminium or cement), nearly doubling since 2000 (ibid).

Plastics are also produced from natural gas, feedstocks derived from natural gas processing and crude oil refining. Plastics can also be made from ethane, an abundant by-product of the gas extracted through hydraulic fracturing, or fracking as it is more commonly known. With an abundance of ethane flooding the market, and in the United States in particular, the petrochemical industry is racing to build plants, called ethane crackers, that can be used to form the building blocks of plastics (Leber, 2020). As Figure 4.1 indicates, industry experts project a 40 per cent increase in plastics production over the next decade – meaning plastics could soon move from a sideshow of the energy industry to accounting for more than 20 per cent of global oil consumption, and increasing to nearly 50 per cent by 2050 (Ellen MacArthur Foundation, 2020; IEA, 2018; O'Neill, 2019).[4]

Indeed, in a report focused on oil industry growth, ExxonMobil Corporation executives assured their shareholders that any losses from the renewable energy and electric vehicle transformation would be offset by growth in petrochemicals (ExxonMobil, 2019). Accordingly, several major oil and gas companies have invested a cumulative US$180 billion in future plastics production (Greenpeace, 2020). ExxonMobil is spending US$20 billion on chemical and refining plants across the Gulf Coast; Royal Dutch Shell PLC is building infrastructure in Pennsylvania that will churn ethane

Figure 4.1: Global primary energy consumption by fossil fuel source measured in terawatt-hours (TWh)

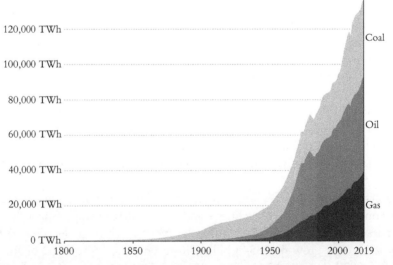

Source: Vaclav Smil (2016). *Energy Transitions: Global and National Perspectives, Second Edition*. Praeger; *BP Statistical Review of World Energy* OurWorldinData.org/fossil-fuels/ (CC BY 4.0)

into 1.6 million tonnes of polyethylene plastic each year; and global oil giant Saudi Aramco aims to invest US$15 billion in a unit of Reliance Industries Ltd that includes one of the largest polypropylene businesses in the world (Shell, nd; ExxonMobil, 2019; Esposito, 2019).

In addition to the US fracking boom, much of the growth in both plastics consumption and production is coming from China and other parts of Asia where economies are quickly catching up with western levels of plastics production and consumption; swiftly turning from world leaders in plastics imports to a major source for plastics production and export in only a few years (Rapoza, 2021). For instance, China-owned Sinopec Corporation has opened three petrochemical facilities in the last two years alone; meanwhile, ExxonMobil has begun constructing a US$10 billion petrochemical complex in Huizhou, China (ibid). Not to be left out, Russia's petrochemical giant Sibur is also heavily invested in the future of petrochemicals, partnering in 2019 with Sinopec to begin constructing the Amur Gas Chemical Complex that is set to become the world's largest basic polymer production facility by 2024 (Woodroof, 2021).

So, in fact, the oil and gas industry is investing billions into *expanding* plastics production. And it is no surprise that the oil and gas industry had the most representatives of any group at the COP26. According to the Centre for International Environmental Law, the plastics boom is a 'classic case of supply driving demand, not the other way around' (CIEL, 2020). Now, amid record low oil prices due to COVID-19 and the shale gas boom, the American Chemistry Council recently proclaimed that the plastics industry has 'never done better business' (Leber, 2020: np). The petrochemical industry has become so competitive that in 2020 BP PLC agreed to sell their petrochemical business for US$5 billion – citing overlaps in their business plan, threats of a global plastics ban and an outright inability to compete with other oil giants that have been building their petrochemical divisions specifically to diversify *through* the energy transition by increasing plastics production (Perkins et al, 2020). So, at this stage, as various bans are declared, implemented and/or delayed due to the COVID-19 pandemic, plastics export routes shift, and oil and gas companies remain entirely focused on profit increases, we may well speculate that our global future is one of increased plastics production and, *ipso facto*, plastics waste.

The situation in Canada may provide insight into these shifts and their potential consequences for plastics waste generation. The Athabasca oil sands, known colloquially as the Athabasca tar sands or the Alberta oil sands, refers to bitumen deposits located in north-eastern Alberta. This deposit – containing an estimated 1.7 trillion barrels of bitumen – sits under some 141,000 square kilometres of boreal forest and muskeg (peat bogs) and is the largest known bitumen deposit in the world (only Saudi Arabia and Venezuela have comparable bitumen deposits). Like petroleum, bitumen is made up of hydrocarbons, but unlike petroleum, at room temperature it is highly viscous and requires much more processing to extract usable oil. This heightened processing requirement means that during periods of oil scarcity (for instance, when Saudi Arabia limits its oil production), Alberta's oil sands are profitable. Conversely, when oil prices are relatively low – as they are at the time of writing this chapter – then processing the oil sands is prohibitively expensive. For this reason, the Alberta oil sands have witnessed several 'boom and bust' cycles.

For some years now, indeed, the relatively low cost of oil has led to a financial crisis in Alberta. As well as gutting funding for education and social programmes as part of the conservative provincial government's 'austerity measures', it is scrambling to secure a viable way to maintain its commitment to oil production. Similar to other regions around the world that have heavily depended on the industrial extraction of oil, gas or minerals, such as miners in the Appalachian mountains (Britton-Purdy, 2016), Albertans are heavily invested in their identity as one of the world's most important oil producers. The clothing company that calls itself 'Alberta Strong', for instance, provides some indication of the strength of the public's pride in its oil industry. Its popular T-shirts and hoodies proclaim 'where the oil flows and families grow' and 'Canadian Oil Kicks Ass'. (As a further sign-of-the-times, Alberta Strong is currently advertising that all of its clothing stock is on 'clearance'). Alberta has always bragged about being the only province to not have a provincial sales tax, and that it is a 'giver' rather than 'taker' in Canadian federal funding redistribution to other provinces. Indeed, the Wild Rose political party of Alberta supports the province's independence from Canada based on its financial ascendancy through oil.

But with the continued drop in oil prices, refusal from its neighbouring province British Columbia to allow Alberta's pipelines to pass through British Columbia territory and the United States' recent stop to the Keystone pipeline, Alberta is on the precipice of moving from being in just another 'bust' part of a 'boom and bust' cycle to permanently losing its financial hegemony. In response, the Alberta government recently released its 'Recovery Plan' (2020a). With an almost exclusive focus on economic recovery (and no more than a nod to environmental recovery or sustainability), Alberta remains focused on "accelerat[ing] the future of the natural gas industry, and a retooled program to incentivize potentially tens of billions of dollars of investment in the petrochemical sector' (2020a: np). The government hopes to establish Alberta – which is already Canada's largest hub for refining and petrochemicals production – among the world's top ten petrochemical producers by 2030, which also happens to be an important year for achieving zero-plastics waste on a national level, according to the 'Canada-wide strategy on zero plastic waste: phase 1 and phase 2'. Alberta's plan includes a ten-year programme meant to attract multi-billion-dollar investments that can only confirm the federal government's commitment to doing whatever it takes to get key oil and gas pipelines built in coming years.

According to Bob Masterson, president and CEO of the Chemical Industry Association of Canada, Nova Chemicals, Inter Pipeline and Canada's Kuwait Petrochemical Corporation are investing CAD$135 million in financial support to build plastics recycling infrastructure. We may speculate, then, that Alberta is making significant plastics recycling infrastructure investments in order to become Canada's recycling hub. To wit, US exports of plastics machinery shipments to Canada and Mexico increased 19 per cent between 2019 and 2020 (Canadian Plastics, 2020).[5]

And here is the framing sleight-of-hand: Alberta is advertising itself as the solution to Canada's plastics waste crisis. As Alberta's 'Natural gas vision and strategy report' argues: 'It is clear the problem is not plastics. It is plastics waste. As Canada's largest and most technologically advanced plastics manufacturer, Alberta is well positioned to demonstrate domestic and global leadership to reduce waste through plastics recycling and repurposing' (Government of Alberta 2022: 15).

Thus, the Alberta government is advertising its investment in new recycling infrastructure as serving the public interest. What it does not make clear is that plastics recycling does not in any way reduce plastics production: plastics recycling focuses solely on post-consumption plastics and therefore consumer responsibility (see Chapter 2). Moreover, plastics recycling requires an *increase* in oil production. That is, increasing plastics recycling necessarily means increasing oil and gas production. As Chapter 3 details, there are numerous problems with recycling: from the fact that very few plastics (about 20 per cent) can actually *be* recycled, to the (often hazardous) waste produced through, and the intense energy requirements of, chemical recycling processing – recycling plastics is certainly not the panacea that industry and government media campaigns have suggested (Rollinson and Oladejo, 2020; MacBride, 2012). Moreover, and key to the subterfuge of the oil and gas industry's frame, plastics can only materially be recycled once (or at most twice) before the quality of the plastics is degraded sufficiently to render it inviable. And to just emphasize this lynch-pin point, the reason the plastics, oil and gas industries are so intimately connected is because plastics production *requires* virgin resin, which is oil. So, increasing plastics recycling means increasing oil production. As Zhu observes, when demand for petroleum drops, companies ramp up their plastics production, and dirty oil production is disguised as green plastics recycling (2021: np). And we have the *appearance* of moving from a linear plastics economy (plastics as waste) to a circular plastics economy (plastics recycling).

In the meantime, Canada continues to be the highest volume importer of US plastics waste (Staub, 2021), and the United States is heavily invested in the profit it secures from importing Canada's plastics waste. It may be for this reason that Canada quietly entered into a three-year bilateral trade agreement with the United States to continue to export waste from Canada to the United States. Canada may have entered into this agreement in order to avoid being sued by the United States under the Investor-State Dispute Settlement process and the North American Free Trade Agreement 2.0. If this sounds far-fetched, there is legal precedent for this. In 1998, in S.D. Myers Inc v. Government of Canada, S.D. Myers Incorporated, a company based in Ohio that transports, processes and disposes of PCB waste, sued the Canadian government on the basis that:

> Canada's export ban on PCB waste between 1995 and 1997 completely undermined its ability to do business in Canada. It claimed that Canada's motive behind the export ban was not concern for human and environmental health, but the protection of Canadian PCB remediation businesses which would not have been able to compete with its business model. SDMI claimed that the export ban breached Canada's obligations under the NAFTA.

The Government of Canada lost the case, and the tribunal awarded CAD$6.9 million to S.D. Myers in damages and costs in 2002. As of the writing of this chapter, the Canadian federal government is considering an amendment to the Canadian Environmental Protection Act that would list plastics under Schedule 1, defining plastics as a toxic substance. This would have powerful effects on regulating plastics exports and handling. A consortium of 63 plastics, chemical and other manufacturing companies and corporations is hard at work lobbying the government in opposition to this amendment, using the now all-too-familiar arguments that the problem is not plastics production but plastics waste; that US plastics exports are the 'feedstock for the Canadian recycling industry'; and that it would violate trade rights between the US and Canada (US Chamber of Commerce, 2020; see also Fawcett-Atkinson, 2020, 2021). Indeed, this consortium is now pursuing legal action against the Canadian government in an attempt to prevent Canada from defining plastics as toxic materials (Environmental Defence, 2021).

Conclusions: framing plastics recycling as an environmental good

As this chapter has demonstrated, the oil and gas industries have largely produced the narrative that plastics are essential to our lifestyles and that the problem is not plastics production but rather plastics waste, and that this problem may be effectively resolved through increased plastics recycling (for which individuals are held responsible). So, on one hand, the plastics industry still touts plastics as a dizzying and indispensable technological advancement. We fly in airplanes composed mostly of plastic parts; we are mobile for longer thanks to plastic hips and knees. Indeed, modern surgery and health care in general is unimaginable without the ubiquitous use of plastics. And on the other hand, these industries are able to transparently increase oil, gas and petrochemical production by non-transparently presenting plastics recycling as an environmentally sound solution. Thus, despite the mounting scientific literature and non-governmental organization reports that plastics have permeated the world's land and oceans, oil and gas industries still quite openly announce that plastics production will continue to increase. And despite this very recent focus on reducing plastics waste through the banning (that is, reduction) of single-use plastics products, recycling plastics remains high on government agendas.

As Kate O'Neill argues, 'it is difficult to intervene in national and corporate decisions, especially when plastics manufacturing corporations are among the most powerful on the planet' (2019: 161). To wit, several municipalities in the Canadian province of British Columbia attempted to ban plastic bags, only to lose in court to the Canadian Plastic Bag Association (a lobby group for the Chemical Industry Association of Canada's Plastics Division), who successfully argued that banning plastic bags would infringe on their companies' profit.[6] Similarly, the American Progressive Bag Alliance started a reverse bag-ban in California in 2015, and several other US states have followed suit, pre-empting municipal authorities from introducing these bans. Plastics recycling companies, along with oil and gas companies, have successfully narrowed the terms of debate to plastics waste. As Chapter 2 demonstrates, the Keep America Beautiful campaign and others like it have proven highly successful in restricting the discussion to litter, for which consumers are then held responsible. If only consumers would correctly sort their plastics, then the problem would be resolved! The plastics recycling industry appears to offer a closed-loop circular system whereby plastics may be endlessly recycled.

And it now appears that plastics recycling is a new resource frontier. Giant multinational waste corporations such as Waste Management Inc and Veolia are increasingly investing in industrial-scale recycling and energy generation. If the speculations in this chapter prove correct, Canada's signing on to the Bilateral Agreement with the United States is a means of avoiding legal action while it buys time for Alberta to create the infrastructure to become Canada's recycling hub. To the extent that Canada will then be managing its own plastics waste rather than exporting it to the United States or developing countries, this is perhaps an environmental 'win' within a larger 'lose'. If the UN creates an international agreement to limit plastics emissions, then Canada will be able to claim that it is meeting this requirement by managing plastics waste 'in house' rather than increasing emissions through plastics waste and recycling exports. Plastics recycling does not decrease plastics, as reusing or (better yet) reducing does, but it *does* maintain oil and gas production, and it does *appear* to offer a circular economy, while actually maintaining a business-as-usual linear economy.

In her analysis of plastics scrap, O'Neill argues that the trade in plastics comes down to an 'either-or' situation/trade-off:

> Trade and recycling of plastic scrap can help us out of the plastics crisis. Preventing this trade might, on the other hand, help wean us off an environmentally damaging product. However ... stopping this trade might be impossible and/or undesirable. For as long as we produce and use plastics, global circulation of scrap matters for building and maintaining a global circular economy. (2019: 151)

Except that plastics recycling does not contribute to the circular economy: it maintains the linear economy. It is for this reason that oil and gas corporations around the world, from Canada to China, are so heavily investing in plastics recycling infrastructure. And, as we have seen from the previous chapters is the case for all types of waste, the focus on post-consumption plastics waste masks the colossal pre-consumption waste and pollution produced in plastics manufacturing.

The Public Problem of PPE Waste and Being Prepared

Introduction

I click on the local online news, and the annoying advertisement pop-up is about toilet paper. Before, I would have muted the sound, but now I watch the advertisement. Charmin, the advertiser, shows a family of bears sitting together watching television. The voiceover claims that Charmin is working hard to ensure that we all have enough toilet paper to make it through the COVID-19 pandemic. And I find myself asking: *do* I have enough toilet paper?[1]

Before the COVID-19 pandemic, environmental tipping points, global climate change, political instability, dwindling primary resources, the ever-widening gap between wealthy and poor, mass migration and exponential capitalist growth – among other critical realities – were enough to furnish myriad imagined apocalyptic futures. Now that we are experiencing successive waves of the COVID-19 pandemic, an apocalyptic future has mainstreamed to millions of people whose economic and political privilege has hitherto cocooned us from the lived experiences of environmental collapse. And while the international COVID-19 response by-line might be 'We're all in this together', our daily news feed suggests otherwise.

As the World Health Organization urges a coordinated global effort to prevent the further spread of COVID-19 and the IPCC and other supra-governmental bodies remain focused on developing systems of resilience and adaptation for populations vulnerable to the effects of climate change, a disparate group of people self-defined as 'preppers' have already been mobilizing themselves and their loved ones for myriad imagined apocalyptic scenarios. Once represented by social media as individuals on the margins of mainstream society if not mental stability, the COVID-19 pandemic is positioning prepping – if not preppers themselves – as a rational and, indeed, responsible answer to the pandemic. Preppers distinguish themselves from 'hoarders' based on their purposeful stockpiling of things anticipated to be vital to surviving short- or long-term disaster. Whereas hoarding is characterized by the volume of disorganized clutter of a jumble of things with no direct survival value, preppers systematically purchase and organize things for survival under anticipated imminent and/or future conditions of extreme societal collapse. But nor are preppers part of a growing 'preventer' identity and movement: preppers aim to protect themselves and their loved ones *after* disaster strikes; preventers focus their actions on precluding disaster from taking place. This includes small everyday actions designed to lower environmental footprints at the individual level, such as reusing and repairing clothing and small appliances, dumpster-diving and scavenging from other people's discards to much more community-engaged actions such as civil protest, government lobbying and so on.

This chapter focuses on prepping as a particular response to the uncertainty of our species' survival. Drawing on a range of theoretical traditions and empirical observations, I critically examine the various discourses and practices that preppers deploy in preparing themselves and their loved ones for what they believe is the probability, if not certainty, of a survivalist future. Far from the experiences of millions of people who have already been forced into relentless adaptation due to unremitting poverty, inequality and global changes in climate, preppers largely plan for their imagined future by accumulating survivalist skills and *things*. That is, what most characterizes preppers is their heightened consumption: preppers spend many thousands of dollars – sometimes millions of dollars – stockpiling stuff: weapons, 'bug-out' gear, non-perishable food, water, clothing, generators and more, sometimes in secret bunkers or other hidden places. Walmart, Costco and other mainstream consumer havens offer emergency food storage kits, and publishing and distributing companies sell prepping guides for adults and children alike. Not only, then, do they eschew initiatives that seek to prevent an imagined future apocalypse, as preventers do, but as this chapter will detail, preppers influence the very conditions that they then feel compelled to respond to as they intensify the hegemony of over-consumption (and its rather more silent yet far bigger twin, over-production). As such, this chapter argues that the increasingly popular phenomenon of prepping is a contemporary reiteration of western consumer/trashing culture, which feeds the global neoliberal capitalist system responsible for the very apocalyptic conditions to which preppers believe they must respond.

Anchoring this analysis are two interrelated questions familiar to waste studies scholars that concern the dialectical relationship between the individual and society. First, how do we avoid simply characterizing prepping as an individual behaviour, and the phenomenon of prepping as the aggregate of these individual actions? That is, how do we conceptualize prepping less as a psychological coping mechanism and more as a *socius logy*, or logic of society? This is critical if we are to refute the individualization-of-waste frame, discussed in Chapter 2. And, second, how do we consider the related problem of scale (Liboiron, 2013, 2014; Hird, 2021) whereby preppers are held more accountable for the environmental costs of their comparatively small-scale accumulation of consumer products and the waste it produces than environmental costs (including waste) accrued through the extraction, production and distribution of products that are – by orders of magnitude – greater?

Si vis pacem, para bellum (If you want peace, prepare for war)

While demonstrating to his sons how to shoot guns, self-proclaimed 'prepping entrepreneur' Tim Ralston shot himself in the thumb. Appearing on National Geographic's *Doomsday Preppers*, a reality television series that aired from 2012 to 2014, in many ways Ralston reifies the prepper stereotype: he is American, white, middle-aged, male and appears fond of reciting National Rifle Association mantras such as 'every American household should have a gun. Never have enough guns, never enough ammo' (Doomsday Preppers, 2012: S1, E3). This particular reality show participant's apocalyptic scenario of choice is a nuclear disaster, which causes an electro-magnetic pulse that at least temporarily (and perhaps permanently) disables America's energy grid and knocks out its transportation and communications systems. Also corresponding well with the prepper stereotype, Ralston began his prepper lifestyle in secret, hiding from his family the approximate US$30,000 he purchased in weapons, ammunition, food and other supplies. Commenting, with relief, about his firearm accident, Ralston told the interviewers "thank God it wasn't my son" (ibid).

During its run, the reality TV show *Doomsday Preppers* featured weekly snapshots of individuals preparing for either cataclysmic natural or human-made disasters. The show may have provided insights into the lived experiences of people preparing for disaster, but its focus is nothing new. Numerous films, some based on novels, depict human-made environmental disasters (*The Day After Tomorrow, Children of Men, Mad Max, The Road, On the Beach, Threads, Miracle Mile* and *Dr Strangelove*); natural disasters (*Sunshine* and *Melancholia*); zombie apocalypse (*World War Z*); global alien invasion (*War of the Worlds, Edge of Tomorrow* and *Signs*); robot ascension (*The Terminator* and *I, Robot*); or – yes, global viral infection (*The Seventh Seal, The Omega Man, 12 Monkeys, Outbreak* and *Contagion*) or some combination of these. Of course, the mainstreaming of disaster and apocalypse stretches much further back in time than these films. The so-called 'first doom boom' occurred during the Cold War, as Americans in particular (but also Soviets - see Brown, 2013) worried about nuclear war and its aftermath. Further back still, the Christian New Testament's 'Book of Revelation', also known as the 'Apocalypse of

John', details the 'apokalypsis' or unveiling – the reckoning of all humanity with the forces of good and evil. We find variations on the end-of-days theme in other major religions such as Islam (The Hour), Judaism (Day of the Lord, War of Gog and Magog) and Hinduism (Vishnu's return to earth in the form of Kulki to destroy the forces of evil). Max Brooks, author of *The Zombie Survival Guide: Complete Protection from the Living Dead*, nicely enjoins the contemporary agnostic with the ancient religious: 'If you believe you can accomplish everything by "cramming" at the eleventh hour, by all means, don't lift a finger now. But you may think twice about beginning to build your ark once it has already started raining' (2003: 159).

As Indigenous scholars and activists (for instance Watt-Cloutier, 2015; Hoover, 2017), environmental racism researchers (for example Nixon, 2011; Adeola, 2012) and, more recently, supra-national organizations such as the IPCC (2020) have pointed out, *real* apocalyptic conditions arrived hundreds of years ago for Indigenous peoples violently subjugated by genocidal colonization; Black and other racialized people enchained through slavery and subsequent discriminatory laws, policies and practices (such as South Africa's Apartheid, Canada's Indian Act and the United States' Jim Crow); people forced to migrate due to the effects of global warming, and poor people devastated by inequitable labour systems, the World Bank, the IMF and so on. The UNHCR, for instance, 'estimates that there are over 65.6 million forcibly displaced people worldwide, of whom approximately two-thirds are internally displaced and therefore unprotected by international law ... [with another] 10 million stateless individuals' (UNHCR 2016 in Alexander and Sanchez, 2020: 11). These are people who live in a state of unrelenting emergency. One third of the global human population have no electricity. Moreover, it is clear that those most responsible for climate change are not living with its already-here effects: the poorest 45 per cent of human beings are responsible for just 7 per cent of anthropogenic carbon emissions while the richest 7 per cent create 50 per cent of the world's carbon emissions (Malm and Hornborg, 2014).

But these past and present unnamed millions are not preppers; they are *actually* living with chronic disasters that have shifted into permanently degraded living conditions, in the here and now. Media and research studies most frequently depict preppers (for instance Foster, 2014; Klein, 2008; Mills, 2018; Robbins and Moore, 2013; Foster, 2014) as motivated by the *anticipation* of future disaster, or an apocalypse yet to come. Preppers fear a diverse range of impending disasters, from solar flares that wipe out electrical grids to terrorist attacks to peak-oil to cataclysmic floods, that more closely resemble something out of a movie rather than the already-here apocalyptic living conditions of the world's abject (Kristeva, 1980; Agamben, 1995; Schneider-Mayerson, 2013). A series of polls – taken pre-COVID-19, and summarized by Michael Mills – shows that US residents, in particular, have a host of anxieties: '[O]ver 40% of Americans fear losing a loved one to terrorism … international conflict (47.5%), economic collapse (44.4%), cyber-attacks (39.1%), a collapse of the electrical grid (35.7%) and biological warfare (41.8%)' (2018: 3). Research estimates that (pre-COVID-19) there were between three to five million Americans who identified with the prepping movement (ibid). Of course, these are vague estimates because a critical feature of the prepper is that they prepare for disaster in secret in order to prevent other people from stealing their things, a point I return to later in the chapter.

And while media and researchers report on prepping as a phenomenon found in a range of countries and cultures in the globalized north, the most popular representation of the prepper is the white middle-class American man, woman and/or family because the United States is most strongly associated with the quintessential 'personality trait' of preppers: consumerism with an emphasis on buying guns and ammunition. Indeed, prepping is a *consumer bonanza*. There are the lazy-person's ready-made non-perishable prepping staples: for instance, TheReadyStore.com sells the READY pre-2000 Food Storage Supply Kit for a 27-year life span for US$3,683.25. Filtration and cleaning supplies may be purchased from the Berkey Light System for US$200–300. Wise Food Storage sells a one-year supply of food for two people for US$2,595 and Costco's Chef's Banquet All-Purpose Readiness Kit sells a 20-year supply of 600 meals for one person for US$149.99. Caro shipping containers may also be purchased for US$1,000 each in order to store all of these purchased products.

For the more committed prepper, there is a cornucopia of 'essential' products to purchase. Bug-out gear includes predictable things like generators, candles and hand-crank emergency radios, knives, guns and ammunition but also less obvious things like gold and silver. The list is, by design, endless. As Gwendolyn Foster highlights, 'there is a lot of money to be made out [of] prepping' (2014: 16). To wit, the revenue for some dehydrated and preserved food companies increased by as much as 708 per cent in 2007/8 (Murphy, 2013). There are also books to be bought: *A Complete Beginner's Guide to Prepping* (Tactical.com, nd), *The Prepping Guide: SHTF Plans* (Brown, nd), *The Prepper's Water Survival Guide* (Luther, 2015), *When Technology Fails: A Manual for Self-Reliance, Sustainability, and Surviving the Long Emergency* (Stein, 2008), *The Prepper's Blueprint* (Pennington and Luther, 2014), *When All Hell Breaks Loose: Stuff You Need to Survive When Disaster Strikes* (Lundin, 2007) and many more. All in all, and before the COVID-19 pandemic, prepping was a US$500 million per year industry (Kelly, 2016). This sum doesn't include Larry Hall's Survival Condo bunker, which alone cost some US$20 million (Garrett, 2020).

And since preppers live in a constant state of preparation (there are never enough guns and never enough ammo), they lament that it is only their finances that constrain further consumption: it's "all I can afford" says Meegan Hurwitt (Doomsday Preppers, 2012: S1, E1), "right now [fuel is] too expensive" says Dennis Evers (ibid: S1, E2) and there is simply "not enough space in the apartment" says Jason Charles (ibid: S1, E3), leading preppers to use whatever space is available until, eventually, those who can afford it move to bigger houses, purchase bigger secret bunkers and/or more storage containers. Indeed, Charmaine Eddy's analysis suggests that 'ideals of ... property ownership and proper object consumption' centrally occupy the prepper's ambitions (2014: 2). More stuff needs more space and, as Eddy points out, in this regard it can be a fine line between prepping and hoarding. Guns and ammunition are particularly popular prepper purchases. Pat and Lynette Brabble, for instance, a retired American couple in their 60s and featured on *Doomsday Preppers*, strongly tether their religion, gun ownership and the American Constitution together. As Pat remarks from their secret stockpiled room, they thank God "for all the provision[s] that he's [God] made [for them]", which includes over 100 guns (Doomsday Preppers, 2012: S1, E3). Indeed, the experts on the show who assess preppers' provisions often counsel participants to purchase and store more products, especially guns, and to carry their guns with them at all times, "for your protection".

All of this consumption takes place within the context of a global neoliberal capitalist system dependent on ever-increasing extraction, manufacturing, distribution, retail, consumption and, inevitably, waste. And managing all of this waste is big business: not only is haulage and disposal a multibillion-dollar industry but so is the remediation of superfund and other contaminated waste legacy sites (Hird, 2021; Beckett and Keeling, 2019). Indeed, the global production of waste and its contamination of human bodies, wildlife, land and water, itself constitutes one of the key dystopian futures that preppers seek to shield themselves and their loved ones from. And thus, prepping through hyper consumption accelerates the very environmental condition that they are preparing themselves for: environmental degradation through accelerating industrial and military extraction and production, and its corollary, increasingly toxic and contaminating waste polluting the environment. Like the Terminator who creates the conditions of its own existence, preppers invoke their own environmental apocalypse. Or as Franklin Ginn astutely observes, 'fantasies of apocalypse are both a product and a producer of the Anthropocene' (2015: 352). And now, with the COVID-19 pandemic, prepping has escaped its containment as an identity on the margins of normal society. Prepping is mainstreaming.

Properly masked and drenched in Purell

It is a common children's fable: one little pig who put pleasure before responsibility, a second little pig who simply did not do enough, and a third little pig who was sufficiently prepared to not only save himself but magnanimously save his two unprepared brothers as well. Jacob Riha, who helped me write this chapter, wondered whether he was, in effect, like one of the lazy brothers after running out of toilet paper during the first COVID-19 wave. As COVID-19 began sweeping the globe and preppers everywhere settled into a feeling of vindication, many of us hitherto non-preppers were left stockpiling *stuff* in hopes that consumption might translate into resilience.

At the time of writing, the novel coronavirus (SARS-Cov-2), the agent of COVID-19, has infected more than 251 million people across 190 countries, and caused over five million deaths (Fauci et al, 2020; Johns Hopkins, 2020: WHO, nd). The world's most privileged economies, health care systems and, indeed, governments, are on the verge of collapse. Extraordinary measures designed to stop the spread of COVID-19 (social distancing, border closures, curfews, PPE regulations) have been variably enforced by governments around the world, leaving some 7 billion people scrambling to meet uncertain and shifting economic, social, behavioural and political conditions (Sohrabi et al, 2020). By March 2020, much of the world was forced into lockdown as a result of the outbreak, meaning most consumers were following stay-at-home orders and restricted to only leaving their homes for 'essential items'. Like animals preparing for winter, we were told to "stockpile food and medication in [our] homes" (Canadian Health Minister Patty Hadju, *National Post* 26 February 2020 [Bharti, 2020]). At the advice of our governments and media alike, we were urged to purchase "extra stores of things like toilet paper, pet food and feminine hygiene products" (Collie, 2020: np). Panic buying set in as retailers and business owners around the world were forced to limit the number of product purchases like hand sanitizer, bleach and cold medicine. Suddenly, once exclusively the obsession of fringe survivalists, disaster preparedness has become a national pastime: toilet paper is now protected by security guards (Toh, 2020) and we are 'drowning in Purell'[2] (Hesse and Zak, 2020: np). While before COVID-19, a modest three to five million Americans and fewer Canadians were concertedly prepping for future disasters, since the COVID-19 pandemic began, some 52 million Americans and 14 million Canadians have begun stockpiling food and other 'necessities' (Laycock and Binsted, 2020).

Disasters typically influence both production and consumption patterns (Larson and Shin, 2018; Pantano et al, 2020; Sheu and Kuo, 2020). For instance, during World War II, industrial productivity increased by 96 per cent (Nelson, 1991). Since the onset of the COVID-19 pandemic, people across the world have displayed stockpiling behaviours that differ in amount and product types from their usual shopping habits – incited by a fear of being like the unprepared and lazy pig brothers who simply did not do enough (Vigdor, 2020). Sales of household cleaning and disinfectant products have dramatically increased since the World Health Organization declared a global health emergency. According to Statistics Canada, for instance, year-over-year sales growth of hand sanitizer, masks and various cleaning supplies reached as high as 639 per cent, 404 per cent and 180 per cent, respectively; while paper towels and toilet paper increased by 288 per cent during the same period (2020; Bedford, 2020). In the United States, the same pattern occurred: panic-induced purchasing increased sales of aerosol disinfectants (385 per cent), multipurpose cleaners (148 per cent) and tissue products (60 per cent) (Conway, 2020). Similar stockpiling of grocery items saw household spending in Canada increase by roughly 46 per cent in the spring of 2020 – led most notably by rice (239 per cent), dried pasta (205 per cent), canned vegetables (180 per cent) and flour (179 per cent) (Statistics Canada, 2020). Alongside this, fresh produce sales decreased by 15 per cent while pre-packaged foods and frozen items – wrapped in plastics packaging – surged by 31 per cent (Roe, Bender and Qi, 2020).

As we continue to see more retailers offering emergency kits and survivalist materials (Costco, Walmart and so on), the disaster prep industry has been experiencing a massive sales surge from customers concerned about the risks associated with COVID-19. Hazmat suits, tactical gear, water purification tablets and air filters are just a few of the survival goods sold on TheEpicenter – an emergency gear and food supply website that has stopped picking up the phone because it cannot keep up with the sudden influx of orders (Popken, 2020). A similar company, LHB Industries, boasts about sales of emergency supplies soaring by between 200 per cent and 1,700 per cent during the first few months of the pandemic (ibid). The coronavirus has also caused a surge in firearm sales: just twelve days after former President Trump announced a national emergency (12 March 2020), gun sales in the United States rose from 80,000 per day to more than 176,000 per day – initiating what would grow to become the three highest single months of gun purchases in US history and surpassing 9/11 and the Columbine massacre by more than 12 per cent (Collins and Yaffe-Bellany, 2020). In addition, many of these are first-time gun purchases.

Years from now, in what we all hope will be a post-pandemic time, the most defining images of the coronavirus will certainly include exhausted health care workers, body bags piled in makeshift mortuaries and single-use masks washed up on beaches. The COVID-19 crisis has propelled a rapid expansion in the production of plastics products (masks, gloves, body bags and so on), with governments, hospitals, residential care facilities, schools and so on competing to boost their stockpiles. Meanwhile, everyday consumers are left fighting for their share of supplies, and overwhelmed nations like the United States cease exports of PPE products to Canada and Latin America despite 'significant humanitarian implications' (BBC, 2020: np). Globally, we are on pace to waste more than 129 billion face masks and 65 billion gloves

(Silva et al, 2020), undoing years of work attempting to address the global issue of plastics pollution and demand for fossil fuel derivatives (Adyel, 2020). According to reports from Wuhan, the epicentre of the COVID-19 outbreak, the increased demand for medical supplies produced more than 240 tons of single-use plastics waste *per day* (Eroglu, 2020). The US is producing an entire year's worth of medical waste in just two months (Silva et al, 2020). Adding to this, individual choices during lockdown have seen take-out meals and pre-packaged groceries grow the plastics packaging market from US$909.2 billion in 2019 to US$1,012.6 billion in 2021 (ibid). Collectively, the global health crisis has put immense pressure on waste management systems, posing major environmental challenges to MSW, institutional (health care) and hazardous biomedical waste management.

Before the pandemic, an estimated two billion people worldwide lacked access to waste collection, while 3 billion people lacked access to proper waste disposal (Sarkodie and Owusu, 2020). Now, the impact on this already burdened industry has significantly amplified. In some areas, like New York City, commercial and industrial waste has decreased by roughly 50 per cent (Kulkarni and Anantharama, 2020). However, those same areas have seen residential solid waste generation increase by 5 to 30 per cent, with the total volume of waste in the United States peaking nationally at 20 per cent higher than normal in April 2020 (ibid). As social distancing has led to increases in online shopping and takeout services, a plethora of plastics-wrapped products being delivered to homes has only added to the enormous influx of plastics waste. For instance, the small island city-state of Singapore discarded an additional 1,470 tons of plastics waste from takeout packaging alone during just an eight-week period of lockdown measures (UNCTAD, 2020). Add to this an overwhelming production of medical waste, and those parts of the world hit hardest by Covid-19 might see an unprecedented increase in waste by upwards of 445 per cent (Kulkarni and Anantharama, 2020; Sarkodie and Owusu, 2020). Collectively, the unrestrainable production of waste as a result of COVID-19 has caused illegal dumping to increase by an estimated 300 per cent during lockdown (Sarkodie and Owusu, 2020). Historical data tell us that more than 75 per cent of COVID-related waste will end up mismanaged, filling landfills and city streets and floating in oceans for decades to come (UNCTAD, 2020).

I've already been in this movie

In *Shopping Our Way to Safety: How We Changed from Protecting the Environment to Protecting Ourselves*, Andrew Szasz persuasively argues that the relationship between individualism, capitalism and consumerism effects a pronounced and insidious consequence – the decline of collective action, or what he calls 'political anaesthesia' (2007: 195). Szasz's starting point is the well-traversed risk theory literature (Beck, 1992) that argues that modernity is characterized by a significant increase in people's understanding that they live (and die) with not simply an increase in both the numbers and diversity of potentially harmful hazards but also an appreciation that risks are inherent to our technologically dependent and driven society:

> [I]ndoor air is more toxic than outdoor air. That is because many household cleaning products and many contemporary home furnishings – carpets, drapes, the fabrics that cover sofas and easy chairs, furniture made of particle board – outgas toxic volatile organic chemicals. Ok, we will go outside – only to inhale diesel exhaust, particulates suspended in the air, molecules of toxic chemicals wafting from factory smokestacks. (2007: 1)

With little difficulty, we may add a growing number of risks to this list: COVID-19 and other viruses and bacterial infections and diseases (for instance, salmonella-infected onions recently occupied my local news), heavy metals, endocrine disrupting chemicals, phthalates, herbicides, pesticides and various gases including methane, carbon dioxide, carbon monoxide, hydrogen, oxygen, nitrogen and hydrogen sulphide. There are over seven million known chemicals – 80,000 of which are in commercial circulation with another 1,000 new chemicals being introduced each year (Wynne, 1987: 48). Add to this the approximate 14,000 food additives and contaminants added to landfills from discarded food. The greatest source of environmental cadmium is thought to be from batteries thrown out in domestic waste. Municipal waste incinerators are also known to emit dioxins and furans with their aerial discharges, possibly to worse levels than toxic waste incinerators (ibid; see also Hird, 2021).

For Ulrich Beck and other risk theory scholars, risk is perceived, filtered and made sense of through neoliberal capitalism, which emphasizes a market economy, enhanced privatization, an overall decrease in government control in favour of industry control and a general entrepreneurial approach to profit maximization (Crooks, 1993; Foote and Mazzolini, 2012; for general discussions of neoliberal capitalist governmentality, see Burchell et al, 1991; Foucault, 1984, 1988). Neoliberal capitalism has successfully transformed citizens into individuals, and individuals into consumers. And it is as *consumers* that we are supposed to make sense of environmental risk:

> [T]he brutal fact of ontological insecurity always has an ultimate addressee: the recipient of the residual risk of the world risk society is the *individual*. Whatever propels risk and makes it incalculable, whatever provokes the institutional crisis at the level of the governing regime and the markets, shifts the ultimate decision-making responsibility onto

the individuals, who are ultimately left to their own devices with their partial and biased knowledge, with undecidability and multiple layers of uncertainty. This is undoubtedly a powerful source of right-wing radicalism and fundamentalism. (Beck, 2007: 195, italics in original)

What Beck identifies as right-wing radicalism and fundamentalism takes the much-publicized form of the United States Republican Party's explicit climate-change denial (DePryck and Gemenne, 2017), but also much more mundane everyday individual responses found in preparedness behaviours. Of course, survivalism in America has long been associated with conservatism, racism, sexism (women can goods while men shoot guns) and white supremacy movements (Lamy, 1996). Sean Hannity, for instance, recommends prepping to Fox News viewers (Kelly, 2016). Of the myriad forms that individual responses to environmental risk might take, it is primarily in the 'modality of consumer' within neoliberal capitalist societies (Szasz, 2007: 4). As such, much of preparedness takes the form of consuming (and ultimately wasting) products. Some products, as we have seen, are specifically marketed as disaster preparation (such as high-end bunkers for billionaires) while far more products – indeed everything that *can* be purchased – are now marketed as 'general preparedness' for an unknowable future. The COVID-19 crisis has transformed such mundane items as toilet paper, tampons, bottled water and toothpaste into consumer stockpile 'must-haves'. During COVID-19's first wave, as social media in the globalized north overran with posts about which stores still had toilet paper in stock, companies such as Charmin reassured consumers that increased consumption was the most responsible response to the pandemic rather than as citizens within a polity, which would implicate individuals within their community and society.

Thus, within the neoliberal capitalist system, attenuating environmental risk is to be achieved through relentless individual consumption. Each purchase is directed at protecting the individual consumer and her family rather than her community or society as a whole. We witness the enthusiasm, passion and time, attention and energy that consumers devote to determining the 'best' bottled water or the 'best' organic underarm deodorant in the numerous posts on Facebook, Reddit and other online forums devoted to consumer lifestyles. And having purchased the phthalate-free shampoo and Dasani bottled water, the individual demonstrates her care of her-self (Foucault, 1976), care for her family and care for her environment. Millions of privileged individual acts of consumption that attempt to insulate the individual and her loved ones against environmental risk leave behind an increasing number of individuals, households and communities that struggle with issues of safe drinking water, air quality, non-toxic food and so on:

An affluent minority – a savvy and influential minority whose political influence is disproportionately greater than their numbers – buys out of the toxic environment, believes it has taken care of the problem for themselves, and loses further interest in that particular toxic issue. Support for more substantive reform weakens. At best, as with organic foods, the situation will tend toward the creation of a permanent dual market, the larger of which consists of products manufactured in a toxic work environment and that contain toxic ingredients, and whose production and consumption continues to discharge these substances into the environment. (Szasz, 2007: 208)

Companies such as Airinum have rushed in to profit from the new COVID-19 'mask mania' by selling designer masks for US$99 or more. From Airinum's website, we learn that

[w]e take around 20,000 breaths every day and air is essential for our survival. When we breathe poor quality air, it can severely damage our health and contribute to asthma, respiratory diseases, cancer, strokes and even death. In fact, 7 million people die annually as a result of poor air quality, a number far too high. Whether you need protection from toxic air pollution, itchy pollen, to stay away from bacteria or simply want to be at your very best, the Urban Air Mask will empower you to breathe cleaner and healthy air. (Airinum, nd)

Airinum explicitly acknowledges (indeed highlights) the effect of air pollution in killing seven million people per year. But the solution is not to improve air quality – to take the lead in producing less, lobbying governments, supporting affected communities – but to increase consumption – what industry calls 'empowering' the individual privileged enough to afford their designer masks. And, familiarly, the privileged few respond accordingly within the neoliberal capitalist script, as these testimonials on Airinum's website attest (Airinum, nd):

I was so pleased with these respirators, they fit perfectly, look great, and actually do the job. After wearing it for about an hour, I decided to take it off, thinking that the bushfire smoke had dissipated. However, as I began to remove the mask, I quickly realised this was not the case, rather *the mask had filtered out the smoke so well that it made me think the air was clear.* (David M., emphasis added)

I wear the mask when walking through the centre of London for half an hour, two times a day. The mask fits well and keeps my nose and mouth covered. It does it's job, I can't smell the pollution or cigarette smoke – which I was really happy with as I have a strong sense of smell! The filter feels substantial, *I feel protected from the polluted environment in built-up areas.* (Emma M., emphasis added)

David need not question the association between climate change and Australia or California's recent forest fires and Emma need not question why London's air is so polluted; they opt out of these concerns by insulating themselves from risk by buying designer masks. Indeed, Gwyneth Paltrow, whose company Goop is valued at US$250 million, sports an Urban Air Mask from the comfort of her private jet in her Instagram post to which she claims to have "already been in this movie", referring to her role in the virus pandemic film *Contagion* (Paltrow 2020).[3] How luxurious to experience the COVID-19 pandemic as a movie role. Paltrow, the Kardashians, other celebrities and elites like the Walton family – who own Walmart, which sells prepping kits – all endorse prepping.

Designer masks and designer hand sanitizer are just the tip of a vast, expensive iceberg of products that the privileged are taking advantage of during this global pandemic. Those wealthy enough are opting to use their private planes or charter jets for 'evacuation flights' out of infected areas and into vacations: 'Avoid coronavirus by flying private. ... Request a quote today!' advertises Southern Jet. Others are opting to spend time on their yachts or rented islands (feat. Kim Kardashian's private island birthday party in October 2020) to isolate themselves from infected shoreline communities. "It totally makes sense", the president of B&B Yacht Charter Jennifer Saia exclaimed, whose family spent its spring vacation on a yacht in the Bahamas instead of their regular villa in Italy: "You're keeping your family contained in a very small, should-be-clean environment. And going from your car to your F.B.O. [private jet terminal] to your private jet right onto the tarmac. And from there, right onto your yacht, and not having to deal with the public" (quoted in Williams and Bromwich, 2020: np).

The environmental gain of fewer middle-class people flying overseas has been offset by the dramatic increase in wealthier people using private jets (Sullivan, 2020; Wagner, 2019). And now, according to a *New York Times* (Williams and Bromwich, 2020) investigation, US health care providers are offering what amounts to private VIP emergency room memberships so that wealthy people may further avoid contact with 'the public' should they require hospital care. Luxury bunker sales are on the rise again, as they were at the outset of the Cold War.

As the next chapter will detail, a burgeoning literature illuminates the relationship between environmental risk and inequity (see for instance Adeola, 2012; Nixon, 2011) and consumption and environmental harm more specifically (see for instance Princen et al, 2002; Dauvergne, 2010). Consumption is a primary neoliberal capitalist mode of environmental risk preparedness, and structural inequity is built into this risk:

> There is no ontology of risk. Risks do not exist independently, like things. Risks are risk conflicts in which there is a world of difference between the decision-makers, who could ultimately avoid the risks, and the involuntary consumers of dangers, who do not have a say in these decisions and onto whom the dangers are shifted as 'unintentional, unseen side effects'. (Beck, 2007: 195)

Risk and privilege are relative: middle-class suburbanites may not control the economic system that, say, defines and decides acceptable levels of DDT administered on crop lands in Somalia (although they may), but they do have the economic power to move themselves and their families to more environmentally safe enclaves, and to further insulate themselves through product purchases that are prohibitively expensive for most people. As the *Doomsday Preppers* television series recounts in episode after episode, many preppers are even prepared to assume the risk of financial ruin in order to stockpile US$100,000 or more on food stocks (see for instance, S1 E2).

Thus, prepping constitutes an increase – often exponentially – in consumption and what Peter Dauvergne (2008) terms its 'shadow': increasing extraction and manufacturing producing environmental degradation, which significantly includes waste. As Szasz points out with regard to the consumption of bottled water: 'To make a bottle full of (more or less) clean water, mountain springs are turned into industrial sites, plastic polymers have to be produced, energy used, hazardous wastes and postconsumer, "solid" wastes generated. The production process transforms nature, pollutes nature. Clean consumption is compatible with, even requires, dirty production' (2007: 197).

Just as we know that most waste is produced during a product's production phase (see Chapters 1 and 2) rather than its post-consumption phase, we also know that factors such as the 'extremely large scale of modern industrial life', economic globalization and economic inequality all contribute to what Jennifer Clapp terms 'waste distancing' or the geographical, economic and political distance placed between consumers and waste (2002: 155). This distancing serves to manage waste in ways that do not disturb – and indeed tends to increase – circuits of mass production, mass consumption and, *ergo*, industry profit (Hawkins, 2006; Kollikkathara et al, 2009; Lynas, 2011).

Global climate change, soil depletion and degradation, biodiversity loss, contaminated water, air pollution are all exactly this: global in their ultimate reach and effects. Environmental racism and inequality research demonstrates that, so far, the most acute and profound effects of environmental degradation are experienced by already disenfranchised millions. In other words, people who prepare for environmental disaster are those who are sufficiently privileged *to be able* to prepare; everyone else is already living (and dying) with environmental disaster. Our contemporary neoliberal capitalist characterization of prepping during the COVID-19 pandemic – defending ourselves and loved ones, accumulating survivalist skills and things – obscures the complex web of interdependence that *produces* the products and services that we consume. That is, ever increasing consumption encourages individuals to believe that they are protected while masking (as double-entendre) the intended and unintended consequences of profit-driven production and consumption in terms of rights-diminished, increasingly unsafe and impoverished labour, community economic, political and cultural devastation, (increasingly toxic) waste generation and so on. As such, prepping to protect ourselves and our loved ones contributes to the very disasters that this behaviour seeks to avoid.

Conclusion: an ounce of prevention is worth a pound of cure

The end of the Cold War may have abated anxieties about nuclear conflict and its radioactive fallout (although even this now feels more fragile thanks to former President Trump's apocalyptic bromance with North Korea) or we may be simply living with additional apocalyptic scenarios. The COVID-19 pandemic presents an opportunity to revisit the phenomenon of prepping from the perspective of living in present disaster rather than the more abstract disaster-to-come anticipated by the traditional prepper. Where once represented as an extreme response to general anxieties by a relatively small and generally white, conservative group of people living at the margins of society, COVID-19 joins climate change, biodiversity loss and other already-here realities that are making preppers out of everyone with the means to purchase and stockpile consumer goods. In short, the COVID-19 pandemic has mainstreamed prepping.

Since the Cold War, the neoliberal capitalist leitmotif to environmental disaster and degradation is *blame and shop, shop and blame.* Patriarch Filaret, head of the Ukrainian Orthodox Church of the Kyiv Patriarchate, publicly stated that the Covid-19 pandemic is "God's punishment for the sins of men and sinfulness of humanity... First of all, I mean same-sex marriage. This is the cause of the coronavirus" (Villareal, 2020: np). After Filaret tested positive for COVID-19, his church defended him, stating, "As the head of the church and as a man, the Patriarch has the freedom to express his views, which are based on morality" and asked the public to "pray for His Holiness Patriarch Filaret, so that the All-Merciful and Almighty Lord God will heal the Patriarch" (ibid). Former President Donald Trump and his supporters blame China and the World Health Organization for the "China virus" while the Chinese government blames the US army for the virus (Winter, 2020). And as this chapter demonstrates, along with the familiar blame-game, we are following the same, and in many cases wildly increased, patterns of consumption: 'eco-doom as consumerist spectacle' as Ginn puts it (2015: 353). By marketing design, we consume to insulate ourselves from each other:

> Community, empathy and logic are not easy to merchandise; they are not profitable to corporations who like to keep us divided and conquered. There is *a lot* of money to be made out [of] prepping. Paranoia sells. Guns and ammo and other prepping gear such as underground bunkers and security, food and water, and so on is an extremely lucrative marketplace. (Foster, 2014: 16, emphasis in original)

Insidiously, these individual acts of consumption (multiplied by millions of people, repeated billions of times), are integral to how neoliberal capitalism defines and structures morality, or our relationship to both our immediate and distant communities:

> In combination with neoliberalism, the individual becomes his own 'moral entrepreneur' and thus holds the fate of civilization in [his] hands. The result is a new 'categorical imperative': act as though the fate of the world depends on your action. Separate your waste, ride a bicycle, use solar energy, etc. The key contradiction which is both obscured and revealed here is that the individual is condemned to individualization and self-responsibility, even vis-a-vis global threats, despite the fact that [he] is severed from the decision contexts which escape [his] influence. (Beck, 2007: 169)

Recall that immediately after the 9/11 attacks, President George W. Bush advised citizens to go out and shop (Leonard, 2010). After the first COVID-19 wave, France's government told its citizens to "go out and start spending" (The Local, 2020: np). Not only are we encouraged to believe that increased consumption will resolve global problems such as disease and terrorist attacks but this individualization of responsibility produces the problem of amplification (see Chapter 2), whereby the extractive and manufacturing industries produce far more waste in producing the products we buy (or that remain stockpiled in warehouses or dumped in open dumps, landfills or incinerators if we do not buy) and individuals voluntarily shoulder the burden of responsibility for the ensuing environmental degradation, including waste. Increases in extraction, production and distribution (far more than consuming) brings peak oil, toxic waste and other global environmental crises closer. As such, what characterizes both pre- and post-COVID-19 prepping is increased industrial production – including weaponry – and increasing waste that is increasingly toxic, increasingly finding its way into land and lakes, rivers, seas and oceans, and into our bodies via waste, such as microplastics. One of prepping's significant ironies is that instead of leading to independence from the neoliberal capitalist industrial complex, prepping depends upon it. Indeed, neoliberal capitalism and the industrial manufacturing complex *created* prepping, and thus cannot be called upon to resolve ever-increasing production and consumption.

Former President Trump's 2020 election rhetoric rehashed the neoliberal capitalist accusation familiar since America's participation in the Cold War: that *any* emphasis on community is 'socialist and anti-American' (Foster, 2014: 16). Yet, we also see a lively counter-narrative in grass-roots movements, such as Black Lives Matter and Idle No More, that are clearly demonstrating that millions of people do not have access to the (privileged) individual response of self-insulation from harm and are instead building and supporting community-level solutions. Various alternative futures, from economic steady-state theories, de-growth initiatives, transition towns, citizen libraries to anti-capitalist youth movements, challenge individual-level neoliberal capitalist responses that are accelerating environmental degradation and disaster (Kallis and March, 2015; Libroiron, 2014). Even some preppers resist their portrayal in the media as hyper consumers. Indeed, as Mills found, prepper concerns are 'overwhelmingly non-apocalyptic' and 'that any worst-case disaster was … going to be temporary' (2018: 6). Out of the 39 preppers in Mills's ethnographic study, all of them stockpiled food for only a few months, and only one couple had a bunker.

Still others identify themselves as preventers rather than preppers. For example, preventers put their energies into adopting renewable energy systems such as solar panels and windmills, as well as building environmental resiliency through permaculture and other Indigenous practices. Preventers reuse, refurbish and repurpose materials, frequenting junkyards and yard sales rather than Costco and Amazon. For some, the Covid-19 crisis has provided an opportunity to build stronger connections with their communities, adapting to limitations of the global supply chain and transitioning to a more sustainable way of living (Bodenheimer and Leidenberger, 2020; Forster et al, 2020). So, whereas preppers largely practice what Szasz terms 'inverted quarantine', aiming to *isolate themselves* from environmental disaster through heightened consumption (and thus more waste production) – a practice that has significantly increased during the COVID-19 pandemic through both voluntary and forced confinement – preventers set their intentions on challenging the conditions that are leading to disaster.

Perhaps the third little pig has it right: not only does he save himself but he works hard to create a robust shelter for his brothers. We might interpret this fable within the traditional prepper context: through hard work and preparation, save yourself and your loved ones from whatever wild calamity might befall you (a wolf that can destroy homes with his breath). But there is a more generous reading: it might be a hard lesson learned (two pigs lose their homes), but the best way to thrive in the world is to prevent disaster *collectively*.

6

A Public Sociology of Waste

Introduction

Wearing masks and keeping our distance from each other and everyone else, I recently walked around the circumference of Frédéric-Back Park with Hillary Predko, one of my graduate students. The walk was months in anticipation, as COVID-19 pandemic restrictions in Montréal have precluded such get-togethers. But on this sunny day, Hillary and I – like the other people we see bicycling, walking their dogs, jogging or picnicking with members of their 'bubble' – are grateful and more than a little elated to be (socially distanced) among other people.

And what a place to be! Frédéric-Back is not just any park. Some 192 hectares (474 acres) in the Villeray-Saint-Michel region of the city, this park is built on top of a landfill. The area's industrial life began as a privately owned quarry, which the Miron family transformed into a landfill when neighbouring residents protested against the constant blasting (Ville de Montréal, nd). The City of Montréal bought the site in 1988 and continued to fill the quarry with MSW. Over the course of the 1990s, the city constructed a recycling centre and a biogas power plant to collect and convert the methane produced by the landfill into energy (linear economy). And as of 2020, nearly 75 hectares of this vast space is still being used for landfilling waste (ibid).

In countries in the globalized north, waste repositories tend to be sited away from residential communities so that (municipal solid) waste is more likely to be out-of-sight and, according to plan, out-of-mind. That is, as Chapters 2 and 3 detail, individuals, families and households are supposed to consider, take responsibility for and shoulder the financial, behavioural and moral burden of waste from the moment of post-consumption (that is, MSW) to the moment of disposal or diversion. After that, we are meant to forget about what happens to our waste. Concerned about the health, environmental and financial effects of living near landfills or incinerators, many people strongly prefer – and lobby for – landfills and incinerators to be as far away from their homes as possible, which often means these waste technologies are sited near communities without the money, power and/ or influence to object. Whether in relatively rich or poor countries, poor and/or racialized communities are far more likely to live on or near waste sites, which may be engineered facilities or open dumps. And these waste sites fill up not only with local waste but with waste that is continuously exported from countries in the globalized north.

As such, Frédéric-Back Park is an altogether different experience. First of all, the whole area is located on the island of Montréal, which is considered 'downtown', and residential homes border the site. There are bicycle paths and pedestrian walkways around the circumference and cutting across the middle area. Benches and picnic tables are dotted around the site, often under new trees planted for shade. There is also a large skateboard park. And at the top of the gentle hill in roughly the centre of the site there sits a garden of shrubs as well as more benches and signs explaining the past and present history of the site as a landfill. All of these features are specifically designed to draw people to the site.

As though this is not enough, an art installation (see Figure 6.1) sits at the top of the hill; the culmination of two years of work by Alain-Martin Richard, entitled *Anamnèse 1+1*. According to the artwork description:

> The two-part work is composed of a rectangular volume made of cast aluminum, the lateral walls of which have been molded from recovered bundles of fabric. From this open container emerges a tree as a symbol of renewal. These walls literally relate the ethnographic thread of the fabric of our lives. The other part of the artwork

Figure 6.1: *Anamnèse 1+1*, Alain-Martin Richard. Frédéric-Back Park

Source: Alain-Martin Richard

includes 30 partially buried calcareous stones sprinkled along the footpaths, a short distance from the central element. Words are engraved on these stones, along with reproductions of residents who took part in the artist's intervention. *Anamnèse 1+1* avoids nostalgia, as it revives the memories and honours the participation of those who are the life's blood of this part of the city. It finds a balance between the living and the inert and transforms the wound in the land into a radiant flowering that points to the future. (Art Public Montréal, 2017: np)

From fast fashion to growing trees: from waste to regeneration. This is beyond utility – this is waste transformed into art, an expression of human creative imagination and beauty. And it is not the only transformation on display. Conspicuously dotting much of the park are white spheres that physically encapsulate the biogas pipes sticking up from the landfill. The locked spheres are intended to protect the biogas pipes while also drawing attention to them. As if these space-age giant white spheres are not attention-grabbing enough, the spheres glow a green colour at dusk. And without a hint of irony, the park boasts that it is 'zero waste'. There are no garbage cans on this transformed landfill: visitors must leave with any waste they create at the site. Of course, in practice this means a visitor will carry their waste to a garbage can outside of the park, and then it will be collected and brought back to the landfill section of the park, all using non-renewable fossil fuels. The irony appears to be entirely unself-conscious.

Frédéric-Back Park has won several awards, including the Environmentally Sustainable Projects Awards' Gold Medal, the International Award for Liveable Communities 2004 Merit, a special mention in the Communities in Bloom 2004 and the Espace Montréal at Expo 2010 in Shanghai. Tourism Montréal boasts that:

[t]he Parc Frédéric-Back ... laid out on the former quarry and city's *dump*, is the most ambitious environmental rehabilitation project ever undertaken in Montréal. Today, thanks to its layout and the array of activities it offers, it is the site of a vast green space dedicated to the environment and culture. (Tourism Montréal, nd: np)

Hillary tells me that the park is named in honour of Frédéric Back, a Québecquois painter, illustrator and film creator. Indeed, Back won an Academy Award in 1988 for his animated short film *The Man Who Planted Trees*. The film depicts a man walking in a desert who encounters an elderly man responsible for planting trees, one-by-one and year-after-year, that enable animals and humans to thrive. No wonder the landfill park is named in Back's honour: this is a story of individual power to transform landscapes, of what one person can do to save the environment. It is the problem of amplification in material, story and symbolic form.

Frédéric-Back Park is not the only waste site that intentionally and explicitly draws attention to itself *as a waste site*. The Central Organisation for Radioactive Waste, or COVRA, boasts that it is the only company that the Netherlands has charged with collecting, processing and storing all of the country's radioactive waste (COVRA, nd). But it is the manner of radioactive waste containment – including high-level radioactive waste – that is truly remarkable. COVRA has built, and manages, a radioactive waste storage facility in Zeeland in the municipality of Borsele in the southern part of the Netherlands. Borsele is also home to a nuclear power station. In what it calls 'the art of preservation', COVRA built Haborg, a storage facility designed by William Verstraeten with Albert Einstein's famous $E = Mc^2$ equation painted on its outside in huge script. The building is currently painted a highly visible bright orange. Every 20 years, starting in 2023, Haborg will be repainted in slightly lighter shades of orange, to represent the decreasing radioactivity, and therefore risk, of the nuclear waste stored within. As COVRA's advertising states: 'The structure's colour will gradually become less intense, in the same way that in time, the heat and radiation produced by the waste stored inside will gradually decrease. The building thereby reflects the concept of radioactive waste in an accessible manner' (COVRA, nd: np).

In 2017, COVRA finished construction on VOG-2, which stores depleted uranium. Also designed by the same artist, the building is 'bright blue with a few orange stripes, and 15-metre long stainless-steel pipes sticking out of the roof edge at three corners of the building. These pipes make VOG-2 the biggest sundial in Europe. This too is a reference to the time factor, which ultimately renders radioactive waste harmless' (ibid). But the spectacle does not end there. COVRA has constructed a third building that contains both low- and intermediate-level radioactive waste along with art and museum pieces that are displayed 'between the containers of radioactive waste' (ibid). Thus, members of the public are *encouraged* to visit this radioactive waste storage site as a combined radioactive waste storage facility, art exhibition, museum and restaurant. Visitors pay money to dine in close proximity to radioactive waste.

In a similar vein, the newly constructed Amager Bakke waste-to-energy facility in Copenhagen, Denmark, is also intended to draw crowds of visitors to re-conceptualize waste as a benign amusement park. Completed in 2017, the plant burns municipal, industrial, construction and institutional waste collected from the region as well as other countries such as the United Kingdom (Murray, 2019). But in a move reminiscent of the 'wait there's more' television steak knife commercials, this waste facility is also a ski slope, a running path, an 80-meter-high artificial climbing wall, a café and a tree grove! As Danish architect Bjarke Ingels, whose company designed the artificial ski slope, enthusiastically claims: "A power plant doesn't have to be some kind of ugly box that blocks the views or casts shadows on its neighbours. It can actually be, maybe the most popular park in the city" (in Murray, 2019: np). Visitors to the plant's rooftop can enjoy views of the city and harbour as well as ski, jog or walk the slope. People live in nearby apartment buildings, and as one local resident exclaimed, "to be totally honest, I don't even think about it – that it's waste" (ibid). And this is largely the intention of these emerging facilities: with the prowess and ingenuity of design, science and technology, waste – even radioactive waste – may be so safely controlled and contained that it may be put to work in creating the material foundation for not just living, but for our play and amusement. These facilities extract, produce, manufacture and retail waste as controlled transformation, as human hubris, for visitors to consume again and again. And, while the materiality of the radioactive or MSW stays the same (or decays very slowly) within a firmly linear economy, the waste-as-lifestyle is presented as part of an innovative circular economy.

Indeed, when visitors walk through the halls looking at the art collection and museum pieces at COVRA's facility in the Netherlands, or are watching bison as they roam the former Rocky Mountain Arsenal plutonium processing plant in Colorado, United States, converted into a national wildlife refuge (Krupar, 2013), they are partaking in what Jean Baudrillard (1983) termed 'simulacra'. Baudrillard theorized four essential stages in the creation of simulacra. In the first stage, a faithful copy is made of something that is meant to 'reflect a profound reality' (ibid: 6). A waste dump reflects the profound reality of mounting waste in all of its stench, largess, obtrusiveness and hazard. In the second stage, images and signs are not understood to reveal reality, as such, but rather intimate some sort of vague and unclear reality that this sign cannot encapsulate. The landfills and incinerators that remove waste from sight, either burying or burning waste, are examples of this second stage. In the third stage, the reality of something has disappeared, and the sign pretends to be a faithful copy of something that has no original. This, for Baudrillard, is the 'order of sorcery' (ibid), whereby meaning is created artificially in reference to an inaccessible truth. We may think of EfW facilities, and indeed zero waste lifestyles here: both create an understanding of *waste that is not*. The final stage is pure simulacrum, whereby the thing bears no relationship to reality, and consumers are not expected to be provided with, or even know or want, anything real. Indeed, any lingering gestures towards reality are discarded as merely sentimental. Amager Bakke's ski slope, café and running track and COVRA's restaurant are simulacra insofar as they bear no memory of waste: they are pure privileged lifestyle. In this sense, these simulacra are the inverse of Georges Bataille's 'accursed share' (1949/91) insofar as waste becomes immanent, rather than excessive, to the economy. In semiotic terms, this is what Chapter 2 refers to as framing waste as not waste.

Frédéric-Back Park, Haborg, VOG-2 and Amager Bakke are also fascinating examples of what we might think of as *inverted sacrifice zones* that dot the global landscape, from New Mexico, United States, to Fukushima, Japan. Sacrifice zones are physical places where military and other industrial activity has sufficiently contaminated the area as to make it permanently uninhabitable and otherwise unusable (Cram, 2015; Krupar, 2013; Masco, 2006; Reno, 2019). These sacrifice zones include both areas whose contamination is a by-product of industrial and/or military activity and areas to which we have intentionally brought contamination.

Such is the case with the Waste Isolation Pilot Plant, about 42 kilometres from Carlsbad, New Mexico, which is storing transuranic nuclear waste for, it is hoped, at least the next million years. This site is cut through with stratifications: the nuclear waste from hospitals, and plutonium from some 30,000 US nuclear weapons is stored in the hole created from the extraction of salt that was produced during the flooding and evaporation of the Permian Sea some 250 million years ago. The Waste Isolation Pilot Project, or WIPP as it became known, concerned itself with how to safely store radioactive waste generated at defence facilities. In 1983, the Department of Energy commissioned a number of anthropologists, archaeologists, linguists and science fiction writers to develop some kind of physical and semiotic warning system to peoples of the future to the radioactive waste stored underground. The Sandia Laboratories recommended that this communication comprise four levels of increasing communication complexity:

Level I: Rudimentary Information: 'Something man[sic]-made is here'
Level II: Cautionary Information: 'Something man[sic]-made is here and it is dangerous'
Level III: Basic Information: Tells what, why, when, where, who and how
Level IV: Complex Information: Highly detailed written records, tables, figures, graphs, maps and diagrams (for a comprehensive examination, see van Wyck, 2005).

Moreover, the message would need to be conveyed in as many written languages as possible, and convey something to the effect of:

This place is a message ... and part of a system of messages ... pay attention to it!
Sending this message was important to us. We considered ourselves to be a powerful culture.
This place is not a place of honor ... no highly esteemed deed is commemorated here ... nothing valued is here.

What is here is dangerous and repulsive to us. This message is a warning about danger.

The danger is in a particular location ... it increases toward a center ... the center of danger is here ... of a particular size and shape, and below us.

The danger is still present, in your time, as it was in ours.

The danger is to the body, and it can kill.

The form of the danger is an emanation of energy.

The danger is unleashed only if you substantially disturb this place physically.

This place is best shunned and left uninhabited.

And in case generations far into the future – after all, the interment must last a million years – no longer recognize the written translations on offer, a series of images of intentionally stark human-constructed landscapes will need to visually capture the same message.

These are public performances – performances *for* the public – meant to engage the public with waste issues in ways far different from that of Frédéric-Back Park, Haborg, VOG-2 and Amager Bakke. Here the message is that waste is dangerous: take note and remember to never come here again. As though the *problem* is not digging it up; not disturbing the waste, rather than the societal structures, economies, politics and norms that led to this highly toxic waste's creation. At Amager Bakke, the message is more like, 'come and play, we have successfully transformed whatever might have been dangerous here into lifestyle'. But, what these two vastly different responses have in common is the same underlying message: through human ingenuity and technological prowess, we (can and do) *manage* waste. And the appropriate public's engagement with these technological–aesthetic fixes is through consumption.

'The trouble with normal is it always gets worse'

To the degree that zero waste behaviours concentrate on reducing, then from an environmental perspective, we are on a very modest right track. Individual zero waste behaviours focused on reduction alone – without community action – are very limited, at best. To the degree that zero waste individual behaviours and/or group initiatives are dependent upon recycling, we are confronted by the myriad environmental costs of recycling detailed in Chapter 3. Equally concerning is the fact that recycling largely amounts to a form of what Samantha MacBride calls 'busyness': a 'sense that progress is being made' even though the facts show that it is not (2012: 171).

Indeed, zero waste initiatives have much in common with the Amager Bakke and COVRA approach to waste. Take Bea Johnson's popular zero waste lifestyle website, blog, 'bulk finder', public speaker, book and – yes – store. According to Book Riot, Johnson's book *Zero Waste Home* is the 'Bible for the zero waste pursuer' (Pereira, 2017). Her online store features everything from shopping kits, products for the kitchen, bathroom, hygiene and beauty, kids and gifting, travel and clothes, office, school and yard. Clicking on each product reveals Johnson's explanation of why she uses the item and where to find it. Most of the items are linked to Amazon.com. For instance, Johnson's online store features a carry-on suitcase, which she states is "[t]he perfect size luggage for my minimalist wardrobe. And it has a[n] unconditional lifetime warranty! Travelling with a carry-on eliminates the need for check-in (unrecyclable) labels, it saves time at the airport and it ensures your luggage won't get lost!"

Clicking on the product details reveals that this piece of luggage (the Briggs & Riley Baseline Softside CX Expandable Carry-On Upright Luggage, Black, 19-Inch) retails for US$499.99 on Amazon. Her book, translated into over 25 languages, also retails at Amazon. Johnson's zero waste enterprise is similar to zero waste initiatives across Europe, which heavily emphasize recycling as an environmentally (and morally) sound solution to waste. Johnson says nothing about the environmental costs of all of the travelling she does to deliver the 'talks in 70 countries on 6 continents, and 23 international speaking tours' that she has already been paid to give. Nor does she remark that her zero waste website encourages consumers to buy more products, and to buy them from one of the world's major polluters, and notorious packaging users – Amazon. Johnson does not comment on the environmental costs of extracting from the earth, manufacturing and retailing (including transporting) the metal drink containers and makeup remover containers she markets on her website. She does not comment on the environmental costs of the products made of wood that she sells. Nor does she comment on the environmental costs of reproducing her print book, nor the environmental costs of her internet use, including her website, blog, Facebook page, Instagram page, Twitter and bank account and so on. Lest we think this is negligible compared with the environmental good that zero waste creates, consider this. The BBC notes that '[t]he European Commission-funded Eureca [project's] lead scientist, Rabih Bashroush, calculated that five billion downloads and streams clocked up by the song Despacito, released in 2017, consumed as much electricity as Chad, Guinea-Bissau, Somalia, Sierra Leone and the Central African Republic put together in a single year' (in Griffiths 2020: np).

As such, we may well think of Zero Waste Home as the 'poster child' for the waste moral economy (see Chapters 1 and 3). Bea Johnson has created an enormously successful business – according to Idol Network, her company is worth US$11 million and she is one of the richest French bloggers – marketing an 'ethical lifestyle'. Zero Waste Home feeds the privileged public's desire to be good environmental citizens, and particularly women's feelings of responsibility to be domestically prepared. As Chapter 5 examined, this is consumption (and in the case of Johnson herself, hyper consumption of energy and carbon emissions from her frequent travel and heavy internet use) justified in the name of saving the environment. Michael Maniates notes that the 'collective obsessing over an array of "green consumption" choices and opportunities to recycle is noisy and vigorous, and thus comes to resemble the foundations of meaningful social action' (2002: 51–2) or 'the dynamic ability of capitalism to commodify dissent' (ibid: 46).

This is not to say that trying to live a little-waste lifestyle is morally bankrupt. To the degree, as Andrew Gilg et al (2005) point out, that these individual behaviours are positively associated with community-organizing and action to hold government and industry accountable for reducing the waste that industry (including the military) overwhelmingly creates, then there may well be a synergy between these two scales. So, for instance, to the degree that an individual attempts to consume as few new products as possible *and* joins forces with community, national and international groups to lobby companies like Amazon to reduce their packaging (as well as pay fair taxes, fair wages, protect their labour force and so on), then individual lifestyle changes are not harmful, and indeed may work in tandem with community- and nation-level involvement. But evidence shows that one (individual lifestyle change) does not necessarily lead to the other (community engagement) and may, indeed, lead to a *decrease* in community or up-scaled action. As Nina Mazar and Chen-Bo Zhong (2010) found in their empirical study, consumption is tightly connected to people's moral and social selves. Indeed, these researchers found that people rated individuals who consumed so-called green products more morally and socially favourably than those who consumed non-green products. At the same time, people who consumed green products were not only less likely to share their money but were also significantly more likely to lie and steal. This is what is known as the 'licensing effect': people 'are least likely to scrutinize the moral implications of their behaviors and to regulate their behaviors right after their moral self has experienced a boost from a good deed. This implies that virtuous acts can license subsequent asocial and unethical behaviors' (ibid: 495). In a similar vein, George Monbiot (2009) calls this 'consumer democracy', whereby

some people have more votes than others, and those with the most votes are the least inclined to change a system that has served them so well. A change in consumption habits is seldom effective unless it is backed up by government action. You can give up your car for a bicycle – and fair play to you – but unless the government is simultaneously reducing the available road space, the place you've vacated will just be taken by someone who drives a less efficient car than you would have driven (traffic expands to fill the available road-space). Our power comes from acting as citizens – demanding political change – not acting as consumers.

We may put this more bluntly in the form of a simple equation: Techno-fixes + individual responsibility = environmental degradation and, eventually, the global crisis we are in. And the trouble with this normal state of affairs, as Bruce Cockburn observes, is that it always gets worse.[1]

Towards a public sociology of waste

As long as individuals privileged enough to afford so-called eco-products satisfy themselves that they are meaningfully and sufficiently contributing to protecting the environment, the longer this kind of 'busy work' (to use MacBride's succinct term) will serve extractive, manufacturing and retail industries in deflecting attention from their shocking and environmentally untenable creation of this literal mess, our global waste crisis (and we may add here the related global crises of climate change and biodiversity loss). Let's remember that it was the Container Corporation of America that sponsored the creation of the now-ubiquitous recycling logo. And it was BP Oil that created the individual carbon footprint calculator, effectively deflecting attention away from their – by orders of magnitude – much higher carbon footprint than any individual could create (Kaufman, nd).

In order to meaningfully tackle our local and global waste problems, waste must be understood within the context of overarching upstream critical issues involving historical and ongoing settler colonialism, poverty, and racialized and gendered relations. When discussions of waste are isolated from inequalities, we effectively corroborate and maintain the industrially created and planned focus on individual choices and individual responsibility. In short, waste is a profound and enduring *symptom* of the inequalities of poverty, race and gender, and must be framed as such.

A number of studies are framing waste as a social justice issue, focusing on the association between waste and poverty, for instance where open dumps, landfills and other waste repositories are proposed and sited (for example, Amegah and Jaakkola, 2016; Furedy, 1997; Mothiba et al, 2017; Parizeau, 2006). And far from confined to the globalized south, these open waste sites, and their toxicity, are features of the globalized north. As MacBride and many others point out, 'industrial zones – the only suitable spots for large-scale processing of recycling as well as garbage transfer, disposal, and incineration – are overwhelmingly near the homes of people of color and sometimes working-class white people' (2012: 125).

Canada's Indigenous communities are a case in point. Situated in one of the wealthiest countries in the world, and constantly topping the list of countries most desirable to live in, Indigenous communities face shocking and largely unresolved basic life challenges such as safe drinking water and affordable and safe housing. Indeed, a number of Indigenous communities across Canada have endured decades of water advisories and forced plastic water imports because of water polluted by extractive and manufacturing industries, leading some Indigenous communities to launch a class-action lawsuit against the federal government (Cecco, 2021). In Nunavut and the other northern territories of Canada – as in many settler colonial regions across the world – open dumping is the norm. Many of the Arctic's original waste dumps were sited and developed alongside resource exploration sites in the early to mid 20th century, whereby sites were selected primarily for ease of access by short-term non-residents, including, significantly, the Canadian and American militaries (Johnson, 2005); a resulting patchwork of contaminated and ill-designed dumpsites burdens already traumatized communities. These abandoned waste sites are a source of dangerous environmental contaminants. Among them are the Cold War-era radar stations that make up the Distant Early Warning (DEW) Line, around which the soil commonly exceeds the 50 ppm of PCBs allowable under the Canadian Environmental Protection Act (Stow et al, 2005). PCBs are known endocrine disruptors and potential carcinogens. Because PCBs bioaccumulate and bio-magnify in the tissues of wildlife (Braune et al, 1999; Giesy and Kannan, 1998) and can leach distances of up to 25 kilometres (Pier et al, 2003), many Inuit country foods have become sources of contaminant exposure (Van Oostdam et al, 2005).

As such, waste is a particular symptom of (settler) colonialism as well as of poverty, race and gender injustice (Anderson, 2010) where waste is configured as a fallout of neoliberal capitalism – an (un)anticipated supplement – which can be managed as a technological issue (bigger and better waste facilities) and individual responsibility for diversion (primarily recycling). This is why individual-level and forms of what Paul Wapner describes as 'liberal environmentalism' are ineffective:

Liberal environmentalism is *so* compatible with contemporary material and cultural currency that it implicitly supports the very things that it should be criticizing. Its technocratic, scientist, and even economistic character gives credence to a society that measures the quality of life fundamentally in terms of economic growth, control over nature, and the maximization of sheer efficiency in everything we do. By working to show that environmental protection need not compromise these maxims, liberal environmentalism fails to raise deeper issues that more fundamentally engage the dynamics of environmental degradation. (1996: 21–2)

A public sociology of waste, then, needs to take a different approach to resolving our global waste crisis than the current mainstream liberal environmentalist approach that maintains that our planet's main waste problem consists of MSW, and that this waste may be effectively resolved through the twin efforts of increased individual responsibility and technological innovation. This approach – long the favourite of extractive, manufacturing and retail industries – has proven to effectively maintain the status quo in the service of neoliberal capitalist growth. The only *actually* effective approach is one that forefronts waste as a social justice issue, and as such, this needs to be a public sociology of waste's main focus.

Centralizing social justice necessarily engages citizens with its causes: for starters, neoliberal capitalism, the marketization of nature, the power of extractive and manufacturing industries nationally and globally, racism and sexism and the profound global socio-economic inequities between people. According to Oxfam International's '5 shocking facts about extreme global inequality':

> [T]he world's richest one percent of people have more than twice as much wealth as 6.9 billion people; only four cents on every US dollar comes from taxing the wealthy; one out of every five children (some 258 million) do not go to school; about 10,000 people die each day because they have no access to affordable health care; and, men own fifty percent more of the world's wealth than women, and the twenty-two richest men have more wealth than all women in Africa. (2021: np)

In her work on meaningful public engagement, Isabelle Stengers strongly argues that we must turn away from our current forms of public engagement that seek 'the consensual transformation of the "ignorant public" masterworld into the "citizens' masterworld". [It is] an Empty Great Idea. It will not work' (2005: 159–60). Adrian Mackenzie similarly points out that 'any public that is completely identified and defined by pre-given processes and forms falls short of democratic political practice' (2013: 484). For Stengers, any hope exists in the form of objection because it 'rejects the differentiation of ignorant publics and knowing science' (Mackenzie, 2013: 481–2). What Stengers (2005: 160) refers to as 'objecting minorities' produce 'not as their aim but in the very process of their emergence the power to object and to intervene in matters which they discover concern them' (ibid). This is the difference, argues Stengers, between a public concerned with validating and participating, and the 'potentially most interesting possible public of all: a verifying public' (Mackenzie, 2013: 491). Such an objecting minority would have to work against the assimilation of public consultation and democratic engagement with waste management, as it is currently organized and controlled by industry through things like assessment exercises set up by industry with local government cooperation.

As we have seen throughout the book, this form of governance leads to waste management's configuration as a technological issue supported by norms and practices that encourage individual responsibility. That is, waste management is largely structured in neoliberal capitalist terms, as a matter of responding to individual citizens' waste 'needs' through industry-produced technology, rather than a social justice issue. Framing waste management as a technological issue circumscribes discussions to focus on better technologies (longer-serving landfill liners, better ways of disposing of incinerator fly ash and so on) and diversion (primarily recycling), the latter for which members of the public are largely held responsible. Waste management industries operating in tandem with municipal governments increasingly ask members of the public to accede to prescribed assessment exercises that circumscribe the parameters to, for example, discussions of 'end-of-pipe' responses (that is, disposal). Once this key parameter is set in advance, discussions are further circumscribed to decisions on a limited number of sites, technologies, consultation and discussion events, and consultation time frames (Ali, 1999; Coninck et al, 1999; Dodds and Hopwood, 2006; Einsiedel et al, 2001; Healy, 2010; Petts, 1998, 2001). Multinational corporations specializing in waste technology assessment, siting, construction, operations, monitoring, closure and aftercare increasingly manage this framing. With on-site engineers and scientists, networks with government, and sophisticated, well-budgeted, in-house public relations management teams, these new brokers increasingly manage municipal and public discussions of waste management through feasibility reports, town hall meetings, presentations and other forms of consultation (Allen, 2007; Corse, 2012; Marres, 2005a, 2005b; Van de Poel, 2008). Indeed, neoliberal governance enhances industry's monopoly by embedding techniques such as public consultations and feasibility studies within industry's remit. In other words, geo-engineering and economics are the primary discourses through which waste operates.

As such, multinational corporations have become vital allies through which municipalities attempt to turn waste issues into a technical problem governed by a clearly demarcated general will and common good. Like Mackenzie's analysis of public participation in scientific and government discussions of synthetic biology, waste management works in similar ways: it may not be touted, as synthetic biology is, as a 'vehicle for global salvation' (2013: 483), but it is discussed in terms of solving a public need. Working with Stengers's flat rejection of a participating democratically engaged public, Mackenzie points out that a public governed by pre-defined parameters obviates democratic political practice (ibid: 484). In the case of waste then, an objecting public would need to work against the assimilation of public consultation and democratic engagement with waste management, and this is difficult to envision since protests tend to engage with the assessment exercises set up by government and industry. For Stengers (2005: 160), an 'objecting minority' would need to refuse the insidious normality of waste's management as such. This public 'might do more than validate'; it may 'risk saying or doing something different'.

And saying and doing something different is precisely, I argue, what Indigenous communities around the world *are* doing. Numerous examples, past and present, abound. For instance, in order to protect the untouched wetlands and their treaty territory under threat, Anishinaabe peoples in the United States and Canada are protesting against the Line 3 pipeline expansion, which would bring some million barrels of tar sands per day from Alberta, Canada, to Superior, Wisconsin, in the United States (see stopline3. org: nd). We might well learn from the Wıı̀lıı̀deh Yellowknives Dene First Nation, whose land is polluted with the infrastructural and gold processing waste produced by the Giant Mine operations in Canada's Northwest Territories, and who now face living with 237,000 tons of arsenic trioxide. As director France Benoit details in her film *Guardians of Eternity* (2015), the Wıı̀lıı̀deh Yellowknives Dene First Nation is both considering and preparing ways to warn future generations of people about a contamination they did not create but have been encumbered with in perpetuity. The Canadian federal government's response to this toxic waste – 237,000 tons of arsenic trioxide – is to attempt to store it in perpetuity. And as Shiloh Krupar wryly notes, ' "in perpetuity" is ultimately an unadministrable charge' (2013: 142). In other words, it is a form of governance through a 'spectacle of oversight that actually minimizes the act of governing' (ibid).

The Kitchenuhmaykoosib Inninuwug (KI) of North Western Ontario challenged the mining company Platinex's plans for mineral exploration on traditional KI land and water, doing so on the argument that their culture is based on an intimate relationship (involving history and identity) with land (Ariss, 2017; Ariss and Cutfeet, 2011). In 2008, six leaders of the KI community were imprisoned for peacefully protesting resource development on their land. The KI-6, as they became known, were released two months into their sentence. The KI community challenged the Ontario courts, successfully arguing that Platinex's plans violated the community's rights under Treaty 9. As the KI-6 repeatedly reminded government and industry officials, their obligation to protect their environment extended to all things (for instance, the fish in Big Trout Lake) and people, not just the KI community itself (Ariss, 2017; Ariss and Cutfeet, 2011). Similarly, the Qamani'tuaq (Baker Lake) community in the Kivalliq Region of Nunavut challenged French multinational Areva's proposal to construct a uranium mine and store radioactive waste 80 kilometres west of their community. And in the wake of Brazil's Córrego de Feijão iron mine waste tailings pond rupture in 2019, which led to environmental devastation some several hundred kilometres long through five Brazilian states and the death of at least 209 local inhabitants, local residents and activist groups as well as international organizations such as Greenpeace are – despite threats and harassment – attempting to hold Vale Canada Limited accountable.

And lest we think of these as local-, regional- or even national-level struggles, Martín Arboleda reminds us that 'the immanent dynamics that underpin the spaces of extraction of late capitalism are *global in content and national only in form*' (2020: 26, emphasis in original). A public sociology of waste must also be cognizant that it is often those communities that are most devastated by what we might call the 'fast violence' of environmental disaster, and the 'slow violence' (Nixon, 2011) of environmental degradation, that are also on the front-lines of challenging highly enfranchised, and government protected, companies and corporations, and are often the most poorly resourced.

Non-governmental organizations such as Friends of the Earth and Greenpeace often have local chapters that people may join and help in efforts to challenge things like waste exportation from rich to poor countries. These organizations also lobby national and supra-national governments. Governments have long depended on multilateral agreements between countries to either solve or manage problems, such as nuclear arms, human trafficking and the rights of children. Several of these agreements pertain to climate change: the 1992 UN Framework Convention on Climate Change; the 1997 Kyoto Protocol and the 2015 Paris Agreement. The 1987 Montréal Protocol reduced, and in many cases, phased out ozone-depleting materials. At the time of writing, countries are considering a global plastics treaty. Already, as Chapter 4 points out, the Basel Convention on the Control of Transboundary Movements of Hazardous Wastes and Their Disposal is a significant step towards reducing waste exports. The Basel Convention's Annex II, which includes recycling, is a vital next step, and pressure needs to be exerted on those countries, such as Australia, the United States and Canada, that have refused to sign on, or are delaying their participation. Social, political and economic movements like the Green New Deal may be harnessed to institute federal-level government regulations, such as Germany's Packaging Act, known as the VerpackG, which as of 2019 requires all producers – including importers and retailers – to take responsibility for the packaging they produce and/or use by paying a recycling fee. The fee charged depends on the amount and type of packing used, providing an economic incentive to producers and retailers to use the least amount of packaging possible. It is notable that Germany instituted the VerpackG in response to the latest revision of the European Union's 2018 Packaging Waste Directive.

Above all, a sociology of waste must critically examine how rightsholders and stakeholders *frame* waste. As this book has detailed, frames determine how the problems of waste are defined, who and what are held and not held accountable, how these problems are either maintained, increased or resolved, and ultimately how the underlying problems of social, economic and political inequality that necessarily undergird waste issues are either recognized, witnessed and tackled, or deflected and otherwise reproduced.

Appendix

This Appendix provides a detailed account of how placentas are redefined, several times, in order to make them available for scientific research. I include this detailed analysis in order to demonstrate how critically important both ontology and epistemology are to framing waste. As this research shows, scientists define placentas as waste (ontology) in order to secure the availability of placentas for their research. And this classification (epistemology) of human tissue as waste (and then not-waste, and then waste again) goes largely (but not entirely) unchallenged.

Phase 1: placentas are waste

Anthropological, nursing and bioethics literatures have recorded diverse interpretations of placentas across cultures and times (see Yoshizawa's 2013 review). Beliefs that placentas are spirit doubles, omens or medicine, potentiate rituals around disposal, such as burial with significant objects, or consumption, such as placental dehydration and encapsulation. Despite working closely and directly with placentas, some participants were not aware of these practices:

> To tell you the truth, I have not come across any incidence where people treated placenta[s] differently. In my experience, every placenta is considered a throwaway tissue. ... I have never met someone who did some ritual thing with the placenta. ... When the use of something is gone, then you basically don't worry about that thing. So the pregnancy is done, delivery has taken place, you have your own baby, why do you want to worry about [the] placenta? [P15]

> I haven't thought that people had access to the placenta, because you know, in the labour room, it was just discarded properly. You wouldn't really give it to the mother, so I never thought people could do anything with this. ... I have no clue about it. I haven't heard of it. [P25]

Some participants were aware of cultural valuations of placentas, but when specifically questioned about them, mostly minimized the popularity and importance of certain public understandings of placentas as non-waste. As in the extract above, some participants suggested that women consider the placenta during their pregnancy as it pertains to the health of their foetus, or that women who have pregnancy complications are more mindful of its role. However, most participants asserted that pregnant women, mothers and the public at large strictly "do not care" about placentas:

> They [pregnant women] never think about the placenta – ever, ever, even during pregnancy. They're thinking of their baby. [The placenta] is so important during pregnancy. But after that, once it's delivered, the baby has no use for it. The mom has no use for it. So it has fulfilled its role, you know? [P8]

Some participants associated the conceptualization of placentas as not-waste with 'cultural minorities':

> I am a very matter-of-fact person, and I see the placenta as being the biological necessity for mammalian reproduction. ... But if [a mother] were of the cultural minority who wanted to take the placenta away, she should have the opportunity to say "no". So I respect people who have those views. I don't know why they have those views, but that's their business, not mine. I don't have any reverence for the placenta, in that sense. I think it's a very interesting tissue. Very versatile, and I think quite extraordinary. But that's a biological science fascination. [P14]

Here, as with other participants' responses, 'cultural minority' interpretations were contrasted with an apparently unified and uniform scientific definition of placentas as waste. A researcher made a distinction between cultural valuations and how they can be respected "scientifically", suggesting such valuations do not have a place within science proper:

> Following women during their pregnancy and then actually discussing separately as to whether we can have their placenta, and taking into consideration that they might be happy to participate in one part of the study, but they may not want us to have their organ, I find that most women are happy to donate their placentas. But you do occasionally have women who say, "no, I want to take my placenta home and bury it". Or plant a tree on top of it or something like that. And I think that's about as respectful as we can be scientifically in relation to other people's placentas. [P20]

Placenta science research commonly accepts a definition of the placenta based on two assumptions: (1) that placentas are intra-uterine tissues that perform a function during pregnancy alone; and (2) that maternal and foetal needs for the placenta are temporary. As such, these assumptions differentiate non-'natural' meanings of placentas as beyond the interest of science (see Yoshizawa, 2016). This determination effectively precludes a recognition that different and/or postpartum purposes necessitated by women, infants or communities might be important for understanding the placenta in science and medicine. It concomitantly determines the meaning of waste in placenta science: biological waste is a culturally meaningless matter whose life-course has run out.

Numerous scientific articles confirm the assumption that placentas may be used for scientific research because they will otherwise be wasted (see Barachini et al, 2009; Chang et al, 2007; Ilancheran et al, 2009; Muralidhar and Panda, 2000; Scalinci et al, 2011; Barbati et al, 2012; Wang et al, 2008; Wolff et al, 1996; Yen et al, 2005; Yu et al, 2009; Zhang et al, 2006). For instance, Sergio Scalinci et al write that they used the placenta as a source of stem cells in part because of 'the absence of ethical issues, considering that the placenta is a waste product' (2011: 691). The scientific definition of placentas leans upon the classification of placentas as strictly 'natural', biological material, without layers of cultural meaning, as evinced by the following response:

> But when I see a placenta in the lab, it's just a specimen. ... I mean, you watch a cat or a mouse or a dog give birth, and the mother just turns around and eats it. I mean it has no value. ... I know a much broader animal vision of it, just being part of a natural process without needing to take on anything extra. [P18]

Considered together, these excerpts suggest that the placenta scientists who participated in our study believed that the placenta is, in essence, biological matter whose functions and roles end with delivery. With no purpose, disposal is the expected next step at the end of the placenta's natural life-course:

> And you know, this piece of tissue, that's all it is: it's a piece of tissue that otherwise would be thrown in the garbage. [P11]

> I think placenta tissues are fine to take because they're really just going to go and get thrown out. They don't have any other use. [P22]

> In science – I mean, we are scientists – it's considered a throwaway tissue. In our clinic, in IRB's, you have this category: throwaway, residual tissue, or useful tissue. It's considered a residual, throwaway tissue. So you don't sometimes even need IRB approval for getting placenta, because it's considered a throwaway tissue. [P15]

A fascinating repeated discursive thread in our respondents' remarks concerned the claim that nature defines placentas as waste. Participants frequently cited the 'disposal' of placentas by animals as evidence for this claim, and compared disposal of placentas in hospital incinerators as the equivalent practice among humans. For example:

It's a fact of how, well, mammalian reproduction that takes place, isn't it? You have pups and all the placenta[s], you have the foetal membranes and everything else, and that also applies to the bird's eggs. When the chick comes out of the egg, the eggs are thrown away, and of course they try to throw it away at a distance so that the predators are not going to find the nest. ... That may be the reason why we like to discard these things. Not to leave traces when you have a placenta. It's just lying there at the entrance of ... a rabbit hole, say. And if the placentas are thrown on the outside, all the foxes will find it. ... Discarding the placenta is a sort of natural process. [P1]

In this sense, discarding the placenta is figured as a survival strategy inherent to reproduction itself. The participant continued:

And that's why a lot of animals eat the placentas of the birth. Even non-carnivores, like for example ruminants. They don't have the enzymic system to get to digest that meat material. But that's an ecological protection. That's why I think they get rid of all that. For example, when so many ... hide their waste products, when they defecate and so on, they hide it and they bury it just to avoid all traces of having been there. Otherwise it attracts others. ... It just came into my mind, but that's a fact: so you never find placentas in nature. They're all hidden, or buried, or eaten. Because they have to, as a protection for the young that are somewhere. [P1]

Within this logic, it follows that because there are "no placentas in nature", all animals with placentas actively discard them after birthing. Humans, as a biological species, do the same via the incineration of placentas. Significantly, what these excerpts suggest is that anything that is disposed of is waste. This is affirmed in the conviction expressed by P1 above that animals do not eat their placenta for its nutritive properties (which would suggest that placentas are not waste but food), because some animals lack the ability to digest their placentas.

Indeed, participants in our study framed the discarding of placentas as a necessary, natural and logical practice. As such, for participants, defining placentas as waste is not actually a process of defining, since placentas are waste.

Phase 2: placentas are not waste

Now that placenta scientists have (as in phase 1) established that women, their families, society at large and indeed nature itself define birthed placentas as waste, a second discursive phase involves scientists redefining placentas as not-waste, evincing the mobilities of the waste hierarchy. This is evident in placenta science publications. For example, in an article published in the journal *Placenta*, Ornella Parolini writes:

> The placenta should never be seen as a waste material, but in addition to claiming its important role during pregnancy, it should be regarded as a great gift from nature as a source of cells and bioactive molecules for therapeutic applications. ... The placenta may continue to sustain our life even outside of the womb. (2011: S284)

The book *Regenerative Medicine Using Pregnancy-Specific Biological Substances* (Bhattacharya and Stubblefield, 2011) contains only one chapter (among 400 pages; Samuel et al, 2011) addressing ethical issues pertaining strictly to cord blood, but laments the 'massive wastage' of expelled reproductive tissues that could be used for research or therapeutic purposes (Burd and Huang, 2011: 3).

Waste claims are not merely rhetorical; rather, they are frames that provide a conduit for placentas to physically move to laboratory spaces as acceptable objects of scientific study. Placentas, that is, are discursively and materially diverted from waste to resource. And it is not so much that mothers, their families and members of the public at large are 'wrong' in defining placentas as waste, as it is that placenta scientists are able to understand placentas as transformable objects (from waste to not-waste). Thus, this re-classification affords a significant amount of leverage to scientists' unique expertise.

In fact, while some respondents advanced the necessity to secure informed consent, some contended that it is a rather superfluous and unnecessary requirement. Since, these respondents assert, women effectively abandon their placenta after birth (and through this action define the placenta as waste), scientists may recuperate this waste material:

> When it's finished, it's finished. And what I regret is that there are so many barriers stopping people using it for research purposes. The ethics, the formal ethical requirements for researching on the placenta are very stringent, and quite ridiculously so, since it's a throwaway tissue. If it's not used for research, the thing is just destroyed. [P14]

> I don't see any ethical issues, because as I said, so far, throwing it away has not hurt anybody. So by keeping it also, I don't think it's going to hurt anybody. You use it as much as you want, and then if you don't

use it, save as much as you want to save, and then the rest of that can go to the dump. ... I don't know, because that's a complicated issue. Women don't even know where the placenta goes. So how can they not consent? They don't even know, where did it go [sic]? So what's the point of not consenting? [P15]

So, I like the idea that we can just use the tissues since they are just about to be discarded without having to get consent. ... After it has fulfilled its purpose, 99.99% of placentas are discarded, you know? ... If it's going to be used for research, should there be consent? If it's going to be thrown out, no, no consent is necessary for that. [P8]

These views are further justified by claiming that there is no harm caused to anyone by using placentas for scientific research, including compromises to patient privacy, as one of the participants excerpted above explained:

Many institutions do not require approval for placenta[s] because as I said it's a throwaway tissue, so we don't require it. But if you are really going to follow and look at the data, clinical data of the person, anyone to compare, then you need prior IRB approval, consent from the woman. ... The consent is for going into her clinical data. [P15]

Participants sometimes even suggested that asking for consent is an excessive 'generosity' given to patients: "I always didn't have [sic] to have informed consent, but I always just ask the individual [orally] if they were happy with us using their placenta in research. ... It was almost a courtesy to these women. And I would always respect their wishes [P5]."

One participant considered that the necessity of informed consent did a great disservice to women:

I had arguments with our ethics committee for ages. They say "you've got to give the woman at least 24 hours to consider the proposal that you're going to use her placenta". I said "well how can you do that? She's out at home, and then she comes in and labour [sic], and then she delivers her placenta. There is no way to do it." Now if it were doing research on her baby, yes. ... I say, "look, when women have been pregnant and delivered, they're not looking to deliver the placenta and then take a picture of it and put it above their mantelpieces". They're after their baby. The placenta is irrelevant. That's true in our culture, anyway. And therefore, what are you fussing about? ... For a woman, she wants to have a baby. She doesn't want to enter into a binding legal agreement which clearly the hospital thinks they are going to argue about for years and years. She wants to be done with the placenta, have it taken away and destroyed, and get on with her life. [P14]

None of our respondents reflected on the possible conflict of interest in asking for placenta donation after birthing, a time of major distraction for mothers and families during which the possibility of obtaining informed consent may be compromised. This was dismissed because, as one participant (P12) claimed, when asked if they would donate their placenta to science, women "always agree". In this collective discourse, we can see that one ethical praxis – seeking informed consent – is viewed as undermining a greater moral good.

Phase 3: expertise and placenta rehabilitation

The placenta scientists in our study strongly agreed that the placenta is not given due scientific attention. For example, one participant talked about how other scientists and clinicians do not understand the value of the placenta:

I think among scientific people, those who do the very specific reproductive stuff, know the value of [the] placenta. Most of the physicians don't know this. When I give classes for my students in medicine, I have to tell them this: the placenta is not a throw-out. We have to understand the placenta, to study the placenta, to give value to this placenta. Most of the physicians think that placenta is [sic] garbage. ... And I can say, in my hospital, if I don't put my student inside the centre [that is, delivery room or caesarean section operating theatre], they will throw it out. So we still have a lot of work to do to change the mind[s] of the people. And even with the [medical] students. [P13]

This is an advocacy issue for placenta scientists. The placentological literature explicitly highlights other scientists' ignorance of the value of placentas for producing scientific knowledge or therapies. For example, Nadia Badawi et al ask 'Why is the placenta being ignored?' (2000), while in an article entitled 'Remarkable placenta', respected placenta scientist Kurt Benirschke claims that the placenta, 'once richly studied', has become a 'forgotten organ' (1998: 194). The journal *Placenta* repeats this imperative to revisit the history of placenta science in order to revive interest and show how the foundational insights of reproductive science came from studies of the placenta (see, for example Carter, 2011; Carter and Mess, 2010; Lapaire et al, 2007; Pijnenborg and Vercruysse, 2008; Pizzi et al, 2012). Numerous articles decry a lack of placental examination after birth (Baergen, 2007; Benirschke et al, 2006; Salafia and Vintzileos, 1990; Tellefsen and Vogt, 2011). The placenta can even be seen to have been 'discarded' from the medical and scientific curricula, a 'transience' directly linked to how animals discard the placenta:

> And the first thing to go is the placenta, because it's so transient, you know? You have it when you're born. We said the animals were eating them. They're discarded. And so pathology for a physician is the organs that form the individual. And that's what you're going to study for 70 or a hundred years depending on how old you become. You're never going to go back and look at this placenta. So it's the first thing to get dropped out of curricula. [P18]

In sum, there is a sense in which, if the views of others that the placenta is of no value after delivery could be corrected, even more important scientific work could be done with delivered placentas. And further, the very specialization of placenta scientists makes them uniquely qualified to identify placentas as not-waste.

Phase 4: placental research as moral imperative

Scientists argue that they are morally obligated to use this potentially bio-valuable material in experimental contexts. The participants in our study regretted that placentas are moribund after delivery unless they 'rescue' them:

> I think it helps so much, for people to study it. It's no risk for her [the mother] or the baby, so it's stupid if you block this. ... I know in Germany – I don't know in Canada – they cannot use the cord blood in some places, because they have to ask the foetus. You know, because it's from the foetus. ... If you think like this, it's stupid. I think we have to take care. Of course, we have to be ethical, of course we need informed consent. But we cannot be stupid. Like if we will throw out the placenta or the cord blood, we'll do nothing with that, even though we can help a lot of women by studying that. [P13]

Indeed, participants showed excitement in discussing the potentials of such placentas, as exemplified by this participant:

> Like I said, it's big, and nobody wants it. And its discarded, from the human subject's standpoint. Often, it's a discarded tissue so you need minimal IRB involvement. Sometimes it can be exempt. ... You know, we do a dissociation procedure where we isolate the trophoblast cells [cells that form the placenta], and you can get a hundred million cells easily. You can get two hundred million cells sometimes, which is a lot of cells to do your studies with. You can do multiple studies. ...

And it's a primary tissue, meaning its not derived from a tumour or a cell line. It's primary tissue culture. This is very rare. The only [other] thing you can do this with is a blood sample. The blood sample is the next easiest thing to do because many people don't mind giving up a little bit of blood. ... We get the placenta without any intervention at all. ... So, identifying those processes and surgical interventions where there's some discarded tissue is a very good thing, I think. The placenta is a real obvious one from that standpoint. [P4]

This registers the final phase of the process of determining placentas as bio-valuable research products. While placental waste material is defined as ethically neutral, our respondents argued that recuperating placentas as not-waste is a positive ethical behaviour; in fact, scientists have a moral imperative to use placentas in scientific research. The majority of participants reflected upon their motives and goals of studying placentas by invoking an altruistic discourse of science and medicine. Three participants' words are exemplary:

The placentas I get are all from healthy women, and someone's had a baby, and they're happy. And you know, this piece of tissue ... that's all it is. It's a piece of tissue that otherwise would be thrown in the garbage. And I guess plus the added thing is that, you know, again, I can do some good with this. But it's garbage. It's not a living, feeling, thinking tissue. And, you know, I'm next door, and I hear the baby cry, and there's a happy event. [P11]

Actually, it's the beginning of all of our journeys. We all were in the same phase, like we all were a foetus. So, it's always good. And if we come to know from the beginning what is happening, then if there is some problem which can affect our adult life, we can remove it at the beginning. So that's why I think it's very important to learn the placenta biology. And anything related to reproductive health, so we can give and produce a good world with healthy people. That's what motivates me. [P21]

It's so important, and I really, really love the area that I'm in because it can make such a difference. I work with two recurrent miscarriage clinics, and a high-risk pregnancy clinic, and it's wonderful when someone gets pregnant naturally and has a healthy, happy baby. ... I mean, I've even been invited to some [of] the deliveries. You get quite close to some of the women, and holding this beautiful baby at the end and knowing you may have had a part of it – it's a very rewarding field. I'm really at the interface of the clinical and lab-based science. [23]

Interestingly, this moral imperative can sometimes be advanced at the expense of others. At a 2013 workshop on building tissue-sharing networks, an attendee shared her displeasure at having to "fight off the mother who wants to eat it [the placenta] or bury it" (field notes). One clinician shared her "trick" for obtaining consent to take the placentas for scientific purposes from women who had already expressed a desire to take them home: reminding them of the countless women and children who could be helped if only they would donate them, and offering to take and email photographs of the organ. Others in the workshop laughed and nodded their heads in seeming agreement with her frustration with these patients.

Notes

Chapter 1

[1] Burawoy's vision of public sociology – that it must situate itself against neoliberal capitalism (as the overarching structuring force in society, which has organized national and global labour, financial institutions, government priorities, families and so on since the 1930s) – has been the subject of considerable debate and criticism. Some contend that public sociology purports to recognize the lived experiences of all members of the public but nevertheless cleaves to social class as the overarching organizational structure of society, rather than race, gender or sexuality (see, for instance, Stacey, Collins and Glenn in Clawson et al, 2007). Some criticize this vision for overestimating the extent to which sociologists agree on the aims and objectives of public sociology; that not only do sociologists vary widely in their understandings of the goals of their profession but also that many sociologists eschew sociology as a political force with a political (left) agenda in favour of sociology as an academic discipline that contributes empirical evidence to the understanding of the lived experiences of people within the structures of society (see, for instance, Nielsen, 2004), leaving political interpretations to others.

Chapter 4

[1] The Basel Convention does not refer to radioactive waste, which is subject to different regulations.

[2] According to Kate O'Neill, the only WTO dispute over waste trading concerned the EU and Brazil in 2006. In its complaint to the European Commission, the EU charged that Brazil was refusing to accept used tires, while Brazil countered that the volumes of used tires were posing human health threats including fires and disease-bearing mosquitoes. The European Commission found that Brazil was discriminating against a single exporter (the EU) but allowed that Brazil could restrict imports from all tire exporters. See O'Neill (2019: 173).

[3] Similarly, the solution is not for factory workers in the United States to wear diapers so that they can process higher volumes of poultry, but for the US government and industry to reduce the volume of poultry processing, hire sufficient numbers of workers so that people can take bathroom breaks and so on (Oxfam, 2016).

[4] Despite this, the US Energy Information Administration (EIA) is 'unable to determine the specific amounts or origin of the feedstocks that are actually used to manufacture plastics in the United States', suggesting a high degree of flexibility in the feedstock that plastics are made from as well as the plastic products it produces (EIA, 2020).

[5] The petrochemical cluster in south-west Ontario has seen little reinvestment since the 1970s. With the exception of Nova Chemicals, '[t]he industry in Sarnia is much more interconnected than in Western Canada, and there are particular facilities where if one of them is no longer economical, it could take 2–3 others with it', warns Masterson (in

Boswell, 2019: np): 'So we're going to be putting our attention to working with the same stakeholders to bring attention to the need to recapitalize. It has to happen in the next 3–7 years. If it doesn't, we're deeply concerned about the outlook for that cluster.'

[6] Thank you to Kyla Tienhaara for bringing this legal case to my attention. The plastics industry lawsuit ended when the BC government changed the law to allow municipalities in the province to control their own waste issues (Fawcett-Atkinson, 2021).

Chapter 5

[1] I thank Samm Medeiros for her early assistance in reviewing the literature on preppers.

[2] Quote by Nellie Bowles, *New York Times*, 24 April 2020.

[3] Quote by Gwyneth Paltrow via Instagram, 26 February 2020.

Chapter 6

[1] Song from Bruce Cockburn's 1983 album of the same title.

References

Abuelsamid, S. (2021) 'GM to make only electric vehicles by 2035, be carbon neutral by 2040', *Forbes*, [online], 28 January, Available from: https://www.forbes.com/sites/samabuelsamid/2021/01/28/general-motors-comm its-to-being-carbon-neutral-by-2040/?sh=50390b1b6355 [Accessed 13 March 2021].

Ademe (2018) 'Modélisation et evaluation des impacts environnementaux de produits de consommations et biens d'equipement', [online], Available from: https://librairie.ademe.fr/dechets-economie-circulaire/1189-model isation-et-evaluation-des-impacts-environnementaux-de-produits-de-consommation-et-biens-d-equipement.html [Accessed 5 May 2021].

Adeola, F.O. (2012) *Industrial Disasters, Toxic Waste, and Community Impact: Health Effects and Environmental Justice Struggles around the Globe*, London: Lexington Books.

Adyel, T.M. (2020) 'Accumulation of plastic waste during COVID-19', *Science*, 369(6509): 1314–15.

Agamben, G. (1995) *Homo Sacer: Sovereign Power and Bare Life*. Trans. Daniel Heller-Roazen. Stanford, CA: Stanford University Press.

Agar, B.J., Baetz, R.W. and Wilson, B. (2012) 'Fuel consumption, emissions estimation, and emimissions cost estimates using global positioning data', *Journal of Air and Waste Management Association*, 57: 348–54.

Airinum (nd) [online], Available from: https://www.airinum.com/ [Accessed 15 May 2021].

Alexander, C. and Sanchez, A. (2020) *Indeterminacy: Waste, Value, and the Imagination*, New York: Berghahn Books.

Ali, S.H. (1999) 'The search for a landfill site in the risk society', *Canadian Review of Sociology*, 36(1): 1–19.

Allen, B.L. (2007) 'Environmental justice and expert knowledge in the wake of a disaster', *Social Studies of Science*, 37(1): 103–10.

Amegah, A.K. and Jaakkola, J.J. (2016) 'Household air pollution and the sustainable development goals', *Bull World Health Organ*, 94(3): 215–21.

American Chemistry Council (2016) 'Study finds plastics reduce environmental costs by nearly 4 times compared to alternatives', [online], Available from: https://www.americanchemistry.com/Media/PressRele asesTranscripts/ACC-news-releases/Study-Finds-Plastics-Reduce-Enviro nmental-Costs-By-Nearly-4-Times-Compared-to-Alternatives.html [Accessed 25 November 2020].

Anderson, W. (2010) 'Crap on the map, or postcolonial waste', *Postcolonial Studies*, 13(2): 169–78.

Andrady, A.L. and Neal, M.A. (2009) 'Applications and societal benefits of plastics', *Philosophical Transactions of the Royal Society B: Biological Sciences*, 364: 1977–84.

Arboleda, M. (2020) *Planetary Mine: Territories of Extraction under Late Capitalism*, Brooklyn: Verso Books.

Ariss, R. (2017) 'Platinex V Kitchenuhmaykoosib Inninuwug: extraction and the role of law in KI's struggle for self-determination', *Contours*, 7.

Ariss, R. and Cutfeet, J. (2011) 'Kitchenuhmaykoosib Inninuwug First Nation: mining, consultation, reconciliation and law', *Indigenous Law Journal*, 10(1): 1–37.

Arms Control Association (2020) 'The nuclear testing tally', [online], Available from: https://www.armscontrol.org/factsheets/nucleartesttally [Accessed 31 December 2020].

Art Public Ville de Montreal (2017) 'Anamnèse 1+1', [online], Available from: https://artpublic.ville.montreal.qc.ca/oeuvre/anamnese-11/ [Accessed 10 May 2021].

Baarschers, W.H. (1996) *Eco-Facts and Eco-Fiction: Understanding the Environmental Debate*, 1st edn, London: Routledge.

Bahers, J.B. (2021) 'En quoi les politiques locales d'économie circulatire no sont pas des résistances aux régimes métaboliques dominants: le cas des métabolismes urbains de Nantes-Nazaire et Göteborg', Paper presented at the Congrès ABSP-CoSPoF conference, 8 April 2021.

Badawi, N., Kurinczuk, J.J., Keogh, J.M., Chambers, H.M. and Stanley, F. (2000) 'Why is the placenta being ignored?', *Australian and New Zealand Journal of Obstetrics and Gynaecology*, 40: 343–6.

Baergen, R.N. (2007) 'The placenta as witness', *Clinics in Perinatology*, 34: 393–407.

Bank of England (2019) 'Polymer banknotes', [online], Available from: https://webarchive.nationalarchives.gov.uk/20191202203834/ https://www.bankofengland.co.uk/banknotes/polymer-banknotes [Accessed 21 December 2020].

Barachini, S., Trombi, L., Danti, S., D'Alessandro, D., Battolla, B., Legitimo, A., Nesti, C., Mucci, I., D'Acunto, M., Cascone, M.G., Lazzeri, L., Mattii, L., Consolini, R. and Petrini, M. (2009) 'Morpho-functional characterization of human mesenchymal stem cells from umbilical cord blood for potential uses in regenerative medicine', *Stem Cells and Development*, 18: 293–305.

Barbati, A., Mameli, M.G., Sidoni, A. and Di Renzo, G.C. (2012) 'Amniotic membrane: separation of amniotic mesoderm from amniotic epithelium and isolation of their respective mesenchymal stromal and epithelial cells', *Current Protocols in Stem Cell Biology*, 1E: 1–20.

Barbosa, F., Bresciani, G., Graham, P., Nyquist, S. and Yanosek, K. (2020) 'Oil and gas after COVID-19: the day of reckoning or a new age of opportunity?', *McKinsey & Company*, [online], Available from: https://www.mckinsey.com/industries/oil-and-gas/our-insights/oil-and-gas-after-covid-19-the-day-of-reckoning-or-a-new-age-of-opportunity# [Accessed 5 January 2021].

Basel Action Network (2016) 'Secret tracking project finds that your old electronic waste gets exported to developing countries', [online], Available from: https://www.ban.org/news/2016/9/15/secret-tracking-project-finds-that-your-old-electronic-waste-gets-exported-to-developing-countries [Accessed 23 December 2020].

Bataille, G. (1949/91). *The Accursed Share: An Essay on General Economy*, New York: Zone Books.

Baudrillard, J. (1983) *Simulations*, Los Angeles: Semiotext(e).

BBC News (2019) 'Why some countries are shipping back plastic waste', [online], 2 June, Available from: https://www.bbc.com/news/world-48444874 [Accessed 20 May 2021].

BBC News (2020) 'Coronavirus: US "wants 3M to end mask exports to Canada and Latin America"', [online], 3 April, Available from: https://www.bbc.com/news/world-us-canada-52161032 [Accessed 21 May 2021].

Beacock, M. (2012) 'Does eating placenta offer postpartum health benefits?', *British Journal of Midwifery*, 20: 464–9.

Beck, U. (1992) *Risk Society: Towards a New Modernity*, London: SAGE.

Beck, U. (2007) *World at Risk*, Cambridge: Polity Press.

Beckett, C. and Keeling, A. (2019) 'Rethinking remediation: mine reclamation, environmental justice, and relations of care', *Local Environment*, 24(3): 216–30.

Bedford, E. (2020) 'Canadian sales growth of personal care and cleaning products due to coronavirus 2020', *Statista*, [online], Available from: https://www.statista.com/statistics/1110802/coronavirus-growth-in-personal-care-and-cleaning-product-sales-canada/ [Accessed 4 May 2021].

Benirschke, K. (1998) 'Remarkable placenta', *Clinical Anatomy*, 11: 194–205.

Benirschke, K., Kaufmann, P. and Baergen, R.N. (2006) *Pathology of the Human Placenta*, New York: Springer New York.

Bharti, B. (2020) 'Coronavirus updates: stockpile food and meds in case of infection, Canada's health minister says', *National Post*, [online], 27 February, Available from: https://nationalpost.com/news/world/coronavirus-live-updates-who-covid19-covid-19-italy-china-canada-wuhan-deaths [Accessed 4 May 2021].

Bhattacharya, N. and Stubblefield, P. (eds) (2011) *Regenerative Medicine Using Pregnancy-Specific Biological Substances*, London: Springer.

Bodenheimer, M. and Leidenberger, J. (2020) 'COVID-19 as a window of opportunity for sustainability transitions? Narratives and communication strategies beyond the pandemic', *Sustainability: Science, Practice and Policy*, 16(1): 61–6.

Boswell, C. (2019) 'Canadian petrochemicals: investments begin to flow', *Chemical Week*, [online], 19 August, Available from: https://chemweek.com/CW/Document/105460/Canadian-petrochemicals-Investments-begin-to-flow [Accessed 9 March 2021].

Bousso, R. (2020) 'BP wipes up to $17.5 billion from assets with bleaker oil outlook', *Reuters*, [online], 15 June, Available from: https://www.reuters.com/article/us-bp-writeoffs/bp-wipes-up-to-17-5-billion-from-assets-with-bleaker-oil-outlook-idUSKBN23M0QA [Accessed 10 February 2021].

Bowles, N. (2020) 'I used to make fun of silicon preppers. And then I became one', *New York Times*, [online], 24 April, Available from: https://www.nytimes.com/2020/04/24/technology/coronavirus-preppers.html [Accessed 4 May 2021].

Braune, B., Muir, D., DeMarch, B., Gamberg, M., Poole, K., Currie, R., Dodd, M., Duschenko, W., Eamer, J., Elkin, B., Evans, M., Grundy, S., Hebert, C., Johnstone, R., Kidd, K., Koenig, B., Lockhart, L., Marshall, H., Reimer, K., Sanderson, J. and Shutt, L. (1999) 'Spatial and temporal trends of contaminants in Canadian Arctic freshwater terrestrial ecosystems: a review', *Science and the Total Environment*, 230(1–3): 145–207.

Brewer, J.D. (2013) *The Public Value of the Social Sciences*, London: Bloomsbury.

Britton-Purdy, J. (2016) 'The violent remaking of Appalachia', *The Atlantic*, 21 March.

Brooks, M. (2003) *The Zombie Survival Guide: Complete Protection from the Living Dead*, New York: Three Rivers Press.

Brown, B. (nd) 'The prepping guide, SHTF plans', *The Prepping Guide*, [online], Available from: https://thepreppingguide.com/about/ [Accessed 26 October 2020].

Brown, K. (2013) *Plutopia: Nuclear Families, Atomic Cities, and the Great Soviet and American Plutonium Disasters*, Oxford: Oxford University Press.

Browne, H. (2008) 'Curbing my appetite for plastic', *Kingston Whig-Standard*, 20 September, 1–2.

Bulkeley, H., Watson, M., Hudson, R. and Weaver, P. (2005) 'Governing municipal waste: towards a new analytical framework', *Journal of Environmental Policy & Planning*, 7: 1–23.

Bulkeley, H., Watson, M. and Hudson, R. (2007) 'Modes of governing municipal waste', *Environment and Planning A, 39(2)*: 733–53.

Burawoy, M. (2004) 'American Sociological Association Presidential address: for public sociology', *American Sociological Review*, 70(1): 4–28.

Burchell, G., Gordon, C. and Miller, P. (1991) *The Foucault Effect: Studies in Governmentality*, Chicago: University of Chicago Press.

Burd, A. and Huang, L. (2011) 'A massive wastage of the global resources', in N. Bhattacharya and P. Stubblefield (eds) *Regenerative Medicine Using Pregnancy-Specific Biological Substances*, London: Springer, pp 3–8.

Callaghan, H. (2007) 'Birth dirt', in M. Kirkham (ed) *Exploring the Dirty Side of Women's Health*, New York: Routledge, pp 8–25.

Canadian Plastics (2020) 'U.S. plastics industry reports another year of trade surplus', *Canadian Plastics*, [online], 23 October, Available from: https://www.canplastics.com/economy/u-s-plastics-industry-reports-another-year-of-trade-surplus/1003454405/ [Accessed 9 March 2021].

Carter, A.M. (2011) 'Thomas George Lee – implantation and early development of North American rodents', *Placenta*, 32: 8–10.

Carter, A.M. and Mess, A. (2010) 'Hans Strahl's pioneering studies in comparative placentation', *Placenta*, 31: 848–52.

Cavé, J. and Tastevin, Y.P. (2021) 'Zones interdites, zones de sacrifices: penser les résistances aux déjections et aux extractions dans une même trajectoire', Paper presented at the Congrès ABSP-COSPOF, 8 April.

Cecco, L. (2021) 'Dozens of Canada's First Nations lack drinking water: "Unacceptable in a country so rich"', *Guardian*, [online], 30 April, Available from: https://www.theguardian.com/world/2021/apr/30/canada-first-nations-justin-trudeau-drinking-water [Accessed 16 May 2021].

Celik B., Rowe, R.K. and Unlü, K. (2009) 'Effect of vadose zone on the steady-state leakage rates from landfill barrier systems', *Waste Management*, 29(1): 103–9.

Center for Sustainability (2012) 'Problems with current recycling methods', *Aquinas College*, [online], Available from: http://www.centerforsustainability.org/resources.php?category=40&root= [Accessed 20 September 2012].

Chang, C.-M., Kao, C.-L., Chang, Y.-L., Yang, M.-J., Chen, Y.-C., Sung, B.-L. Tsai, T.-H., Chao, K.-C., Chiou, S.-H. and Ku, H.-H. (2007) 'Placenta-derived multipotent stem cells induced to differentiate into insulin-positive cells', *Biochemical and Biophysical Research Communications*, 357: 414–20.

Chavis, B. and Lee, C. (1987) 'Toxic waste and race in the United States: a national report on the racial and socio-economic characteristics of communities with hazardous waste', United Church of Christ's Commission on Racial Justice.

Chertow, M. (2009) 'The ecology of recycling', *UN Chronicle*, 46(3–4): 56–60.

Chong, W.K. and Hermreck, C. (2010) 'Understanding transportation energy and technical metabolism of construction waste recycling', *Resources, Conservation & Recycling*, 54(9): 579–90.

Chung, E. (2019) 'Most styrofoam isn't recycled: here's how 3 startups aim to fix that', *CBC News*, [online], 25 March, Available from: https://www.cbc.ca/news/technology/styrofoam-chemical-recycling-polystyrene-1.5067879 [Accessed 20 April 2021].

CIEL (Centre for International Environmental Law) (2019) 'Plastic & climate: the hidden costs of a plastic planet', *CIEL*, [online], Available from: https://www.ciel.org/wp-content/uploads/2019/05/Plastic-and-Climate-FINAL-2019.pdf [Accessed 7 December 2020].

CIEL (Centre for International Environmental Law) (2020) 'Convention on plastic pollution: toward a new global agreement to address plastic pollution', [online], Available from: https://www.ciel.org/wp-content/uploads/2020/06/Convention-on-Plastic-Pollution-June-2020-Single-Pages.pdf [Accessed 14 May 2021].

Cirino, E. (2017) 'Charles Moore is now a two-time Garbage Patch discoverer (and I can tell you what a Garbage Patch looks like)', *National Geographic Newsroom*, [online], 28 July, Available from: https://blog.nationalgeographic.org/2017/07/28/charles-moore-is-now-a-two-time-garbage-patch-discoverer-and-i-can-tell-you-what-a-garbage-patch-looks-like/ [Accessed 2 January 2021].

City Hall Public Notice (2013) 'City wants to recognize remarkable recyclers', City of Kingston, [online], 18 March, Available at: http://www.cityofkingston.ca/-/city-wants-to-recognize-remarkable-recyclers [Accessed 10 September 2013].

City of Kingston (2013) 'Environment, Infrastructure & Transportation Policies Committee: meeting #04-2013', City of Kingston, [online], 16 April, Available at: http://www.cityofkingston.ca/documents/10180/634489/ EIT_Agenda-0413.pdf/3f5652b3-7460-47c2-a1ea-5a55b24a5c33 [Accessed 3 September 2013].

Clapp, J. (2002) 'The distancing of waste: overconsumption in a global economy', in T. Princen, M. Maniates and K. Conca (eds) *Confronting Consumption*, Cambridge, MA: MIT Press, pp 155–76.

Clark, G. and Clark, A. (2002) 'Common rights to land in England, 1475–1839', *Journal of Economic History*, 61(4): 1009–36.

Collie, M. (2020) 'Yes, you should have a coronavirus emergency kit: here's what to include', *Global News*, [online], 10 April, Available from: https://globalnews.ca/news/6665520/coronavirus-emergency-kit/ [Accessed 6 May 2021].

Collins, K. and Yaffe-Bellany, D. (2020) 'About 2 million guns were sold in the U.S. as virus fears spread', *New York Times*, [online], 1 April, Available from: https://www.nytimes.com/interactive/2020/04/01/business/coronavirus-gun-sales.html [Accessed 6 May 2021].

Collins, S.L., Micheli, F. and Hartt, L. (2000) 'A method to determine rates and patterns of variability in ecological communities', *Oikos*, 91: 285–93.

Conference Board of Canada (2013) 'Environment: municipal waste generation', [online], Available at: http://www.conferenceboard.ca/hcp/details/environment/municipal-waste-generation.aspx [Accessed 23 November 2013].

Coninck, P., Séguin, M., Chornet, E., Laramée, L., Twizeyemariya, A., Abatzoglou, N. and Racine, L. (1999) 'Citizen involvement in waste management: an application of the STOPER model via an informed consensus approach', *Environmental Management*, 23(1): 87–94.

Conway, J. (2020) 'Cleaning product sales growth from the coronavirus in the U.S. in March 2020', *Statista*, [online], 6 January, Available from: https://www.statista.com/statistics/1104333/cleaning-product-sales-growth-from-coronavirus-us/ [Accessed 6 May 2021].

Corse, A. (2012) 'Nature as infrastructure: making and managing the Panama Canal watershed', *Social Studies of Science*, 42(4): 539–63.

COVRA (nd) 'The art of preservation', [online], Available from: https://www.covra.nl/en/radioactive-waste/the-art-of-preservation/ [Accessed 18 May 2021].

Cram, S. (2015) 'Wild and scenic wasteland: conservation politics in the nuclear wilderness', *Environmental Humanities*, 7: 89–105.

Crooks, H. (1993) *Giants of Garbage: The Rise of the Global Waste Industry and the Politics of Pollution Control*, Toronto: James Lorimer & Company.

Crutzen, P.J. and Stoermer, E.F. (2000) 'The Anthropocene', *Global Change Newsletter*, 41 (May): 17–18.

Dauvergne, P. (2008) *The Shadows of Consumption: Consequences for the Global Environment*, Cambridge, MA: MIT Press.

Dannoritzer, C. (2014) *The E-Waste Tragedy* [documentary].

Davies, A.R. (2008) *The Geographies of Garbage Governance: Interventions, Interactions and Outcomes*, Milton Park: Routledge.

Davis, M. (2007) *Planet of Slums*, Brooklyn: Verso Books.

Deloitte (2019) *Economic Study of the Canadian Plastic Industry, Market and Waste Task 5 – Summary Report to Environmental and Climate Change Canada*. Deloitte LLC.

Dent, M. (2020) 'Polymer recycling technologies 2020–2030', *IDTechEx*, [online], Available from: https://www.idtechex.com/en/research-report/polymer-recycling-technologies-2020-2030/760 [Accessed 14 May 2021]

DePryck, K. and Gemenne, F. (2017) 'The denier-in-chief: climate change, science and the election of Donald J. Trump', *Law and Critique*, 24(2): 119–26.

Dias, S.M. (2016) 'Waste pickers and cities', *Environment and Urbanization*, 28(2): 375–90.

Dias, S. and Fernandez, L. (2013) 'Wastepickers: a gendered perspective', in *Powerful Synergies: Gender Equality, Economic Development and Environmental Sustainability*, United Nations Development Programme, pp 153–5.

Dodds, L. and Hopwood, B. (2006) 'BAN waste, environmental justice and citizen participation in policy setting', *Local Environment*, 11(3): 269–86.

Doomsday Preppers, Season 1 (2012) Produced by National Geographic & Sharp Entertainment, United States: National Geographic Channel.

Du Bois, W.E.B., Anderson, E. and Eaton, I. ([1899]1996) *The Philadelphia Negro: A Social Study*, Philadelphia: University of Pennsylvania Press.

Ducros, H. (2019) 'Approaching Waste through Environmental History: An Interview With Thomas Le Roux', *EuropeNow*, [online], Available from: https://www.europenowjournal.org/2019/05/06/approaching-waste-through-environmental-history-an-interview-with-thomas-le-roux/ [Accessed 5 January 2021].

Durant, D. and Johnson, G.F. (eds) (2009) *Nuclear Waste Management in Canada: Critical Issues, Critical Perspectives*, Vancouver: University of British Columbia.

ECCC (Environment and Climate Change Canada) (2018) 'Canadian environmental sustainability indicators: solid waste diversion and disposal', [online], Available from: http://www.publications.gc.ca/site/eng/9.865728/publication.html [Accessed 14 May 2021].

ECCC (Environment and Climate Change Canada) (2019) 'Economic study of the Canadian plastic industry, markets and waste: summary report to Environment and Climate Change Canada', Prepared by Deloitte, [online], Available from: http://publications.gc.ca/collections/collection_2019/eccc/En4-366-1-2019-eng.pdf [Accessed 5 January 2021].

Eddy, C. (2014) 'The art of consumption: capitalist excess and individual psychosis in hoarders', *Canadian Review of American Studies*, 44(1): 1–24.

Edmiston, J. (2010) 'Don't fear the wrapper: users say clear bags no biggie', *Kingston Whig-Standard*, [online], 12 August, Available at: http://www.thewhig.com/2010/08/12/dont-fear-the-wrapper-users-say-clear-bags-no-biggie [Accessed 10 September 2013].

Einsiedel E.F., Jelsøe, E. and Breck, T. (2001) 'Publics at the technology table: the consensus conference in Denmark, Canada, and Australia', *Public Understanding of Science*, 10(1): 83–98.

Eisted, R., Larsen, A.W. and Christensen, T.H. (2009) 'Collection, transfer and transport of waste: accounting of greenhouse gases and global warming contribution', *Waste Management & Research*, 27(8): 738–45.

Ellen McArthur Foundation (2016) 'The new plastics economy: rethinking the future of plastics', [online], 19 January, Available from: https://www.ellen macarthurfoundation.org/assets/downloads/EllenMacArthurFoundation_ TheNewPlasticsEconomy_Pages.pdf [Accessed 7 December 2020].

Ellen MacArthur Foundation (2017) 'A new textiles economy: redesigning fashion's future', [online], November 28, Available from: https://www. ellenmacarthurfoundation.org/publications/a-new-textiles-economy-rede signing-fashions-future [Accessed 7 December 2020].

Entman, R.M. (1993) 'Framing: toward clarification of a fractured paradigm', *Journal of Communication*, 43(4): 51–8.

Environmental Defence (2021) 'Statement on the plastics industry's announcement to pursue legal action against Canada's efforts to control plastic pollution', [online], Available from: https://environmentaldefe nce.ca/2021/05/19/statement-plastics-industrys-announcement-pur sue-legal-action-canadas-efforts-control-plastic-pollution/ [Accessed 5 January 2021].

Eroglu, H. (2020) 'Effects of Covid-19 outbreak on environment and renewable energy sector', *Environment, Development and Sustainability*, 23: 4782–90.

Esposito, F. (2019) 'Saudi Aramco to make big investment in Reliance petrochemicals', *Plastic News*, [online], 13 August, Available from: https:// www.plasticsnews.com/news/saudi-aramco-make-big-investment-relia nce-petrochemicals [Accessed 3 March 2021].

European Commission (2008) 'Directive 2008/98/EC of the European Parliament and of the Council of 19 November 2008 on waste and repealing certain directives' (OJ L 312), Brussels, Belgium: European Commission, 3–30.

ExxonMobil (2019) 'Multi-billion Gulf investment to create tens of thousands of high-paying jobs', [online], 5 February, Available from: https://corpor ate.exxonmobil.com/Locations/United-States/Growing-the-Gulf/20- billion-Gulf-investment-to-create-tens-of-thousands-of-high-paying- jobs#Beaumont [Accessed 9 March 2021].

Fauci, A.S., Lane, H.C. and Redfield, R.R. (2020) 'Covid-19: navigating the uncharted', *New England Journal of Medicine*, 382: 1268–89.

Fawcett-Atkinson, M. (2020) 'U.S. companies threaten to use CUSMA to fight Canada's plastics ban', *National Observer*, [online], 6 November, Available from: https://www.nationalobserver.com/2020/11/06/news/ us-companies-cusma-canada-plastics-ban [Accessed 13 March 2021].

Fawcett-Atkinson, M. (2021) 'The backroom battle between industry, Ottawa and environmentalists over plastics regulation', *National Observer*, [online], 8 March, Available from: https://www.nationalobserver.com/2021/03/08/backroom-battle-between-industry-ottawa-and-environmentalists-over-plastics [Accessed 9 March 2021].

Flower, W. (2016) 'What operation green fence has meant for recycling', *Waste 360*, [online], 11 February, Available from: https://www.waste360.com/business/what-operation-green-fence-has-meant-recycling [Accessed 10 February 2021].

Foote, S. and Mazzolini, E. (eds) (2012) *Histories of the Dustheap: Waste, Material Cultures, Social Justice*, Cambridge, MA: MIT Press.

Forkert, P.-G. (2017) *Fighting Dirty: How a Small Community Took On Big Trash*, Toronto: Between the Lines.

Forster, P.M., Forster, H.I., Evans, M.J., Gidden, M.J., Jones, C.D., Keller, C.A., Lamboll, R.D., Le Quéré, C., Rogelj, J., Rosen, D., Schleussner, C.-F., Richardson, T.B., Smith, C.J. and Turnock, S.T. (2020) 'Current and future global climate impacts resulting from COVID-19', *Nature Climate Change*, 10: 913–19.

Foster, G.A. (2014) *Hoarders, Doomsday Preppers, and the Culture of Apocalypse*, New York: Palgrave Macmillan.

Foucault, M. (1976) *The History of Sexuality: The Care of the Self*, New York: Éditions Gallimard.

Foucault, M. (1984) 'The politics of health in the eighteenth century', in P. Rabinow (ed) *The Foucault Reader*, New York: Pantheon Books, pp 273–89.

Foucault, M. (1988) 'Technologies of the self', in L.H. Martin, H. Gutman and P.H. Hutton (eds) *Technologies of the Self: A Seminar with Michel Foucault*, London: Tavistock Publications, pp 16–49.

Franklin Associates (2018) 'Life cycle impacts of plastic packaging compared to substitutes in the United States and Canada: theoretical substitution analysis', [online], Available from: https://plastics.americanchemistry.com/Reports-and-Publications/LCA-of-Plastic-Packaging-Compared-to-Substitutes.pdf [Accessed 23 December 2020].

Freinkel, S. (2011) *Plastic: A Toxic Love Story,* Boston: Houghton Mifflin Harcourt.

Furedy, C. (1997) 'Socio-environmental initiatives in solid waste management in southern cities: developing international comparisons', *Journal of Public Health* (Bangkok), 27(2): 150–51.

Galison, P. in Kruse, J. (2014) 'Waste-wilderness: a conversation between Peter Galison and Smudge Studio', *Discard Studies*, [online], 26 March, Available from: http://discardstudies.com/2014/03/26/waste-wilderness-a-conversation-between-peter-galison-and-smudge-studio/ [Accessed 6 May 2021].

Garrett, R. (2020) 'Doomsday preppers and the architecture of dread', *Geoforum* (in press).

Geyer, R., Jambeck, J.R. and Law, K.L. (2017) 'Production, use and fate of all plastics ever made', *Science Advances*, 3(7):e1700782.

Gharfalkar, M., Court, R., Campbell, C., Ali, Z. and Hillier, G. (2015) 'Analysis of waste hierarchy in the European waste directive 2008/98/EC.', *Waste Management*, 39: 305–13.

Ghoreishi, O. (2013) 'Canada highest per capita waste producer compared to other developed nations', *Epoch Times*, 30 January.

Gies, E. (2016) 'Landfills have a huge greenhouse gas problem: here's what we can do about it', *Ensia*, [online], 25 October, Available from: https://ensia.com/features/methane-landfills/ [Accessed 21 April 2021].

Giesy, J.P. and Kannan, K. (1998) 'Dioxin-like and non-dioxin-like toxic effects of polychlorinated biphenyls (PCBs): implications for risk assessment', *Critical Reviews in Toxicology*, 28(6): 511–69.

Gilg, A., Barr, S. and Ford, N. (2005) 'Green consumption or sustainable lifestyles? Identifying the sustainable consumer', *Futures*, 37: 481–504.

Ginn, F. (2015) 'When horses won't eat: apocalypse and the Anthropocene', *Annals of the Association of American Geographers*, 105(2): 351–9.

Goffman, E. (1959) *The Presentation of Self in Everyday Life*, New York: Doubleday.

Goffman, E. (1974) *Frame Analysis: An Essay on the Organization of Experience*, Cambridge: Harper Row.

Government of Alberta (2020a) 'Alberta's recovery plan', [online], Available from: https://www.alberta.ca/assets/documents/alberta-recovery-plan.pdf [Accessed 9 March 2021].

Government of Alberta (2022) 'Getting Alberta back to work: natural gas vision and strategy', [online], Available from: https://open.alberta.ca/dataset/988ed6c1-1f17-40b4-ac15-ce5460ba19e2/resource/a7846ac0-a43b-465a-99a5-a5db172286ae/download/energy-getting-alberta-back-to-work-natural-gas-vision-and-strategy-2020.pdf [Accessed 9 March 2021].

Government of Canada (2016) 'Municipal solid waste and greenhouse gases', [online], Available from: https://www.canada.ca/en/environment-climate-change/services/managing-reducing-waste/municipal-solid/greenhouse-gases.html [Accessed 20 May 2021].

Grajeda, T. (2005) 'Disasterologies', *Social Epistemology*, 19(4): 315–19.

Greenpeace Canada (2020) 'Plastic recycling: that's not a thing', [online], Available from: https://www.greenpeace.org/static/planet4-canada-stateless/1d30117a-greenpeacereport_plasticrecyclingthatsnotathing.pdf [Accessed 5 January 2021].

Gregson, N., Metcalfe, A. and Crewe, L. (2007) 'Moving things along: the conduits and practices of divestment in consumption', *Transaction of the Institute of British Geographers*, 32: 187–200.

Gregson, N., Watkins, H. and Calestani, M. (2010) 'Inextinguishable fibres: demolition and the vital materialism of asbestos', *Environment and Planning*, 42(5): 1065–83.

Griffiths, S. (2020) 'Why your internet habits are not as clean as you think', BBC, [online], Availble from: https://www.bbc.com/future/article/20200305-why-your-internet-habits-are-not-as-clean-as-you-think

Hamilton, L.A. and Feit, S. (2019) 'Plastic and climate: the hidden costs of a plastic planet', *CIEL* (May).

Haraway, D.J. (2008) *When Species Meet*, Minneapolis: University of Minnesota Press.

Harris, S. (2015) 'Canadians piling up more garbage than ever before as disosables rule', *CBC News*.

Hawkins, G. (2006) *The Ethics of Waste: How We Relate to Rubbish*, London: Rowman & Littlefield.

Healy, S.A. (2010) 'Facilitating public participation in toxic WM through engaging "the object of politics"', *East Asian Science, Technology and Society: An International Journal*, 4(4): 585–99.

Hesse, M. and Zak, D. (2020) 'The history and mystery of Purell, the most sacred goo of our new era', *Washington Post*, [online], 26 March, Available from: https://www.washingtonpost.com/lifestyle/style/the-power-of-purell-compels-you/2020/03/26/41243960-6dde-11ea-b148-e4ce3fbd85b5_story.html [Accessed 6 May 2021].

Hinde, D. (2020) 'The Swedish recycling revolution', Government of Sweden, [online], 3 November, Available from: https://sweden.se/nature/the-swedish-recycling-revolution/ [Accessed 14 January 2021].

Hird, M.J. (2009) *The Origins of Sociable Life: Evolution after Science Studies*, Houndmills and Basingstoke: Palgrave.

Hird, M.J. (2011) *Sociology of Science: A Critical Canadian Introduction*, Oxford: Oxford University Press.

Hird, M.J. (2021) *Canada's Waste Flows*, Montréal: McGill-Queen's University Press.

Hird, M.J. and Riha, J. (2021) 'Prepping for the [insert here] apocalypse and wasting the future', in Z. Gille and J. Lepawsky (eds) *Handbook of Waste Studies*, New York and London: Routledge, pp 305–21.

Hodgetts, A. (2018) 'Is plastic actually good for the environment?', *Environment Journal*, [online], 16 November, Available from: https://environmentjournal.online/articles/is-plastic-actually-good-for-the-environment/ [Accessed 7 December 2020].

Hoffman, E.M., Curran, A.M., Dulgerian, N., Stockham, R.A. and Eckenrode, B.A. (2009) 'Characterization of the volatile compounds present in the headspace of decomposing human remains', *Forensic Science International*, 186: 6–13.

Hoornweg, D., Bhada-Tata, P and Kennedy, C. (2013) 'Waste production must peak this century', *Nature*, 502: 615–17.

Hoover, E. (2017) *The River Is in Us: Fighting Toxics in a Mohawk Community*, Minneapolis: University of Minnesota Press.

Hosler, D., Burkett, S.L. and Tarkanian, M.J. (1999) 'Prehistoric polymers: rubber processing in ancient Mesoamerica', *Science*, 284: 1988–91.

Hultman, J. and Corvellec, H. (2012) 'The European waste hierarchy: from the sociomateriality of waste to a politics of consumption', *Environment and Planning A: Economy and Space*, 44(10): 2413–27.

Human Rights Watch (2020) 'Lebanon: huge cost of inaction in trash crisis', [online], 9 June, Available from: https://www.hrw.org/news/2020/06/09/lebanon-huge-cost-inaction-trash-crisis [Accessed 1 January 2021].

Hume, T. and Tawfeeq, M. (2016) 'Lebanon: "river of trash" chokes Beirut suburb as city's garbage crisis continues', *CNN*, [online], 25 February, Available from: https://www.cnn.com/2016/02/24/middleeast/lebanon-garbage-crisis-river [Accessed 14 May 2021].

IEA (2018) 'The future of petrochemicals, IEA', [online], Available from: https://www.iea.org/reports/the-future-of-petrochemicals [Accessed 10 February 2021].

IEA (2020a) 'Global energy review 2020, IEA', [online], Available from: https://www.iea.org/reports/global-energy-review-2020 [Accessed 10 February 2021].

IEA (2020b) 'Global EV outlook 2020, IEA', [online], Available from: https://www.iea.org/reports/global-ev-outlook-2020 [Accessed 10 February 2021].

Ilancheran, S., Moodley, Y. and Manuelpillai, U. (2009) 'Human fetal membranes: a source of stem cells for tissue regeneration and repair?', *Placenta*, 30: 2–10.

Insurance Institute for Highway Safety (2016) *Highway Loss Data: General Statistics*, [online], Availble from: http://www.iihs.org/iihs/topics/t/general-statistics/fatalityfacts/overview-of-fatality-facts

Intergovernmental Panel on Climate Change (IPCC) (2020) 'Climate change and land', [online], Available from: https://www.ipcc.ch/site/assets/uploads/sites/4/2020/02/SPM_Updated-Jan20.pdf [Accessed 21 May 2021].

Islam, M.Z. and Rowe, R.K. (2009) 'Permeation of BTEX through unaged and aged HDPE geomembranes', *Journal of Geotechnical and Geoenvironmental Engineering*, 135(8): 1130–40.

ISWA (2016) 'A roadmap for closing waste dumpsites, the world's most polluted places', [online], Available from: https://www.resource-recovery.net/sites/default/files/iswa_dumpsites-roadmap_report.pdf [Accessed 19 May 2021].

Jambeck, J.R., Geyer, R., Wilcox, C, Siegler, T.R., Perryman, M., Andrady, A., Narayan, R. and Law, K.L. (2015) 'Plastic waste inputs from land into the ocean', *Science*, 347(6223): 768–71.

Jarrige, F. and Le Roux, T. (2020) *The Contamination of the Earth: A History of Pollutions in the Industrial Age*, Cambridge, MA: MIT Press.

Jeong, B.Y. (2016) 'Occupational injuries and deaths in domestic waste collecting process', *Human Factors Management*, 26: 608–14.

John, R. (2012) 'City of Kingston to recognize "remarkable recyclers"', *Kingston Herald*, 29 February, 1–2.

Johns Hopkins University & Medicine (2020) 'Covid-19 dashboard by the Center for Systems Science and Engineering (CSSE) at Johns Hopkins University (JHU)', [online], Available at: https://coronavirus.jhu.edu/map.html

Johnson, K. (2005) 'Structure impacts of landfill sites in cold region communities', in *Waste: the Social Context*, Edmonton: May, pp 11–14.

Judah, J.C. (2008) *Buzzards and Butterflies: Human Remains Detection Dogs*, np: Coastal Books.

Kallis, G. and March, H. (2015) 'Imaginaries of hope: the utopianism of degrowth', *Annals of the Association of American Geographers*, 105(2): 360–8.

Kaufman, M. (nd) 'The carbon footprint sham: a "successful, deceptive" PR campaign', [online], *Mashable*, Available from: https://mashable.com/feature/carbon-footprint-pr-campaign-sham/ [Accessed 17 May 2021].

Kavanagh, M. (2020) *Daughters of Uranium*. Lethbridge, AB: Southern Alberta Art Gallery; Calgary, AB: Founders' Gallery at The Military Museums; Calgary, AB: University of Calgary; and Kitchener, ON: Kitchener-Waterloo Art Gallery.

Kaza, S., Yao, L.C., Bhada-Tata, P. and Van Woerden, F. (2018) *What a Waste 2.0: A Global Snapshot of Solid Waste Management*, Washington, DC: World Bank.

Kelly, C. (2016) 'The man-pocalypse: doomsday preppers and the rituals of apocalyptic manhood', *Text Performance Quarterly*, 36 (2–3): 95–114.

Kingston City Council (2013) 'Information report to the near campus neighbourhoods advisory committee', Report # NCN-13–004, 14 August. Available at: http://www.cityofkingston.ca/documents/10180/1885123/NCN_A0313-13004.pdf/c989083a-8b10-41db-a984-e7d94a1b7a53 [Accessed 25 November 2013].

Klein, N. (2008) *The Shock Doctrine: The Rise of Disaster Capitalism*, Toronto: Vintage Canada.

Kollikkathara, N., Feng, H. and Stern, E. (2009) 'A purview of waste management evolution: special emphasis on USA', *Waste Management*, 29(2): 974–85.

Kristeva, J. (1980/2) *Powers of Horror: An Essay on Abjection*, New York: Columbia University Press.

Krupar, S.R. (2013) *Hot Spotter's Report: Military Fables of Toxic Waste*, Minneapolis: University of Minnesota Press.

Kulkarni, B.N. and Anantharama, V. (2020) 'Repercussions of COVID-19 pandemic on municipal solid waste management: challenges and opportunities', *Science of the Total Environment*, 743: 2–8.

Kurdve, M., Shahbazi, S., Wendin, M., Bengtsson, C. and Wiktorsson, M. (2015) 'Waste flow mapping to improve sustainability of waste management: a case study approach', *Journal of Cleaner Production*, 98: 304–15.

Lagus, T.P. (2005) 'Reprocessing of spent nuclear fuel: a policy analysis', University of Minnesota: Washington Internships for Students of Engineering.

Lamy, P. (1996) *Millennium Rage: Survivalists, White Supremacists, and the Doomsday Prophecy*, Springer: New York.

Lapaire, O., Holzgreve, W., Oosterwijk, J.C., Brinkhaus, R. and Bianchi, D.W. (2007) 'Georg Schmorl on trophoblasts in the maternal circulation', *Placenta*, 28: 1–5.

LaPensee, E.W., Tuttle, T.R., Fox, S.R. and Ben-Jonathan, N. (2009) 'Bisphenol A at low nanomolar doses confers chemoresistance in estrogen receptor-α-positive and -negative breast cancer cells', *Environmental Health Perspectives*, 117(2): 175–80.

Larson, L.R. and Shin, H. (2018) 'Fear during natural disaster: its impact on perceptions of shopping convenience and shopping behavior', *Services Marketing Quarterly*, 39(2): 293–309.

Laycock, R. and Binsted, S. (2020) 'Doomsday prepper statistics', *Finder*, [online], 25 May, Available from: https://www.finder.com/ca/doomsday-prepper-statistics [Accessed 6 May 2021].

Leber, R. (2020) 'Fossil fuel companies are counting on plastics to save them', Grist, [online], 8 March, Available from: https://grist.org/climate/fossil-fuel-companies-are-counting-on-plastics-to-save-them/ [Accessed 5 January 2021].

Leggett, J. (2005) *The Empty Tank: Oil, Gas, Hot Air, and the Coming Global Financial Catastrophe*, New York: Random House.

Lemon, B.S. (2002) 'Exploring Latino rituals in birthing: understanding the need to bury the placenta', *AWHONN Lifelines*, 6: 443–5.

Leonard, A. (2011) *The Story of Stuff*, New York: Simon & Schuster.

Lepawsky, J. (2018) *Reassembling Rubbish: Worlding Electronic Waste*, Cambridge, MA: MIT Press.

Le Roux, T. (2016) 'Chemistry and industrial and environmental governance in France, 1770-1830', *History of Science*, 54: 195–222.

Levis, J.W., Barlaz, M.A., Themelis, N.J. and Ulloa, P. (2010) 'Assessment of the state of food waste treatment in the United States and Canada', *Waste Management (New York, N.Y.)*, 30(8–9): 1486–94.

Lewis, D.L. and Willis, D. (2005) *A Small Nation of People: W.E.B. Du Bois and African American Portraits of Progress*, New York: HarperCollins.

Liboiron, M. (2010) 'Recycling as a crisis of meaning', *Etopia: Canadian Journal of Cultural Studies*, 4: 1–9.

Liboiron, M. (2013) 'Modern waste as strategy', *Lo Squaderno: Explorations in Space and Society*, 29: 9–12.

Liboiron, M. (2014) 'Solutions to waste and the problem of scalar mismatches', *Discard Studies*, [online], 10 February, Available from: https://discardstudies.com/2014/02/10/solutions-to-waste-and-the-problem-of-scalar-mismatches/ [Accessed 6 May 2020].

Local, The (2020) 'French told to "go out and start shopping" to relaunch economy', [online], 29 May, Available from: https://www.thelocal.fr/20200529/french-told-go-out-and-start-spending-to-relaunch-the-econo moy?fbclid=IwAR1H1SNc227Sm3LdJ65L4-sPbhkhYBWN8h53TPuC 9LuUChB-K3tvSHUmvbQ [Accessed 6 May 2021].

Lord, R. (2016) 'Plastics and sustainability: a valuation of environmental benefits, costs and opportunities for continuous improvement', *Trucost*, [online], Available from: https://www.plasticpackagingfacts.org/wp-content/uploads/2016/08/ACC-report_July-2016_v4.pdf [Accessed 9 March 2021].

Lougheed, S. (2017) 'Disposing of risk: the biopolitics of recalled food and the (un)making of waste', PhD Thesis, Available from: https://qspace. library.queensu.ca/bitstream/handle/1974/23777/Lougheed_Scott_C_ 201712_PhD.pdf?sequence=3

Lougheed, S., Hird, M.J. and Rowe, R.K. (2016) 'Governing household waste management: an empirical analysis and critique', *Environmental Values*, 25(3): 287–308.

Lundin, C. (2007) *When All Hell Breaks Loose: Stuff You Need to Survive When Disaster Strikes*, Kaysville, UT: Gibbs Smith.

Luther, D. (2015) *The Prepper's Water Survival Guide*, Berkeley: Ulysses Press.

Lynas, M. (2011) *The God Species: How Humans Really Can Save the Planet*, London: Fourth Estate.

Mabee, W. (2021) 'Biden's Keystone XL death sentence requires Canada's oil sector to innovate', *The Conversation*, [online], 21 January, Available from: https://theconversation.com/bidens-keystone-xl-death-sentence-requires-canadas-oil-sector-to-innovate-153615 [Accessed 5 March 2021].

MacBride, S. (2012) *Recycling Reconsidered: The Present Failure and Future Promise of Environmental Action in the United States*, Cambridge, MA: MIT Press.

Mackenzie, A. (2013) 'From validating to verifying: Public appeals in synthetic Biology', *Science as Culture*, 22(4): 476–96.

Malm, A. and Hornborg, A. (2014) 'The geology of mankind? A critique of the Anthropocene narrative', *Anthropocene Review*, 1(1): 62–9.

Maniates, M. (2002) 'Plant a tree, buy a bike, save the world?', in T. Princen, M. Maniates and K. Conca (eds) *Confronting Consumption*, Cambridge, MA: MIT Press, pp 43–66.

Marten, B. and Hicks, A. (2018) 'Expanded polystyrene life cycle analysis literature review: an analysis for different disposal scenarios', *Sustainability*, 11(1): 29–35.

Masco, J. (2006) *The Nuclear Borderlands: The Manhattan Project in post-Cold War New Mexico*, Princeton: Princeton University Press.

Masterson, V. (2020) 'As Canada bans bags and more, this is what's happening with single-use plastics around the world', *World Economic Forum*, [online], 26 October, Available from: https://www.weforum.org/agenda/2020/10/canada-bans-single-use-plastics/ [Accessed 5 January 2021].

Mazar, N. and Zhong, C-B. (2010) 'Do green products make us better people?', *Psychological Science*, 21(4): 494–8.

Meadows, D.H., Randers, J. and Meadows, D.L. (2004) *The Limits to Growth: The 30-Year Update*, White River Junction: Chelsea Green Publishing Co.

Meikle, J. (1995) *American Plastic: A Cultural History*, Chicago: Rutgers University Press.

Melosi, M.V. (2005) *Garbage in the Cities: Refuse, Reform, and the Environment*, rev edn, Pittsburgh: University of Pittsburgh Press.

Mills, M.F. (2018) 'Preparing for the unknown ... unknowns: "doomsday" prepping and disaster risk anxiety in the United States', *Journal of Risk Research*, 22(10): 1267–79.

Monbiot, G. (2009) 'We cannot change the world by changing our buying habits', *Guardian*, [online], 6 November, Available from: https://www.theguardian.com/environment/georgemonbiot/2009/nov/06/green-consumerism [Accessed 18 May 2021].

Mothiba, M., Moja, S. and Loans, C. (2017) 'A review of the working conditions and health status of waste pickers at some landfill sites in the city of Tshwane metropolitan municipality, South Africa', *Advances in Applied Science Research*, 8(3): 90–7.

Muralidhar, R.V. and Panda, T. (1999) 'Useful products from human placenta', *Bioprocess Engineering*, 20: 23–5.

Muralidhar, R.V. and Panda, T. (2000) 'Value based products from human placenta', *Bioprocess Engineering*, 22: 145–8.

Murphy, L. (2017) 'Sweden's strange problem: not enough trash', *Earth911*, [online], 3 January, Available from: https://earth911.com/business-policy/sweden-trash-problem/ [Accessed 6 May 2021].

Murphy, T. (2013) 'Preppers are getting ready for the Barackalypse', *Mother Jones News*, [online], January/February issue, Available from: https://www.motherjones.com/politics/2012/12/preppers-survivalist-doomsday-obama/ [Accessed 6 May 2021].

Murray, A. (2019) 'The incinerator and the ski slope tackling waste', *BBC News*, [online], 4 October, Available from: https://www.bbc.com/news/business-49877318 [Accessed 10 May 2021].

National Geographic (2019) 'Great Pacific garbage patch', [online], 5 July, Available from: https://www.nationalgeographic.org/encyclopedia/great-pacific-garbage-patch/ [Accessed 2 January 2021].

Nelson, R.R. (1991) 'Diffusion of development: post-World War II convergence among advanced industrial nations', *American Economic Review*, 81(2): 271–5.

Nielsen, F. (2004) 'The vacant "we": remarks on public sociology', *Social Forces*, 82(4): 1619–27.

Nixon, R. (2011) *Slow Violence and the Environmentalism of the Poor*, Cambridge, MA: Harvard University Press.

NPR (2019) 'The litter myth', [online], 5 September, Available from: https://www.npr.org/2019/09/04/757539617/the-litter-myth [Accessed 14 May 2021].

Nuclear Energy Agency (2010) *Radioactive Waste in Perspective*, Paris: Organisation for Economic Co-operation and Development.

Ober, W.B. (1979) 'Notes on placentophagy', *Bulletin of the New York Academy of Medicine*, 55: 591–9.

Olmer, N., Comer, B., Roy, B., Mao, X. and Rutherford, D. (2017) 'Greenhouse gas emissions from global shipping, 2013–2015', *International Council on Clean Transportation*, [online], 17 October, Available from: https://theicct.org/publications/GHG-emissions-global-shipping-2013-2015 [Accessed 14 May 2021].

O'Neill, K. (2000) *Waste Trading among Rich Nations: Building a New Theory of Environmental Regulation*, Cambridge, MA: MIT Press.

O'Neill, K. (2019) *Waste*, Oxford: Polity Press.

Ontario Waste Management Association (OMWA) (2016) 'State of waste in Ontario: landfill report', First Annual Landfill Report, Brampton: Ontario Waste Management Association.

Oxfam (2016) 'Poultry companies are denying their workers bathroom breaks: tell Tyson to #GiveThemABreak', [online], Available from: https://www.oxfamamerica.org/explore/stories/no-relief-for-poultry-workers/ [Accessed 18 December 2020].

Oxfam International (2021) '5 shocking facts about extreme global inequality and how to even it up', [online], Available from: https://www.oxfam.org/en/5-shocking-facts-about-extreme-global-inequality-and-how-even-it [Accessed 17 May 2021].

Paltrow, G. (2020) Instagram Post. Instagram gwynethpaltrow.

Pantano, E., Pizzi, G., Scarpi, D. and Dennis, C. (2020) 'Competing during a pandemic? Retailers' ups and downs during the COVID-19 outbreak', *Journal of Business Research*, 116: 209–13.

Parizeau, K. (2006) 'A world of trash: from Canada to Cambodia, waste is a common problem with common solutions', *Alternatives Journal*, 32(1), 16–19.

Parolini, O. (2011) 'From fetal development and beyond: a continued role for placenta in sustaining life?', *Placenta*, 32: S283–S284.

Pellow, D.N. (2009) 'Electronic waste: the "clean industry" exports its trash', in C. Gossart (ed) *Resisting Global Toxics: Transnational Movement for Environmental Justice*, Cambridge, MA: MIT Press, pp 185–224.

Pennington, J. (2016) 'Every minute, one garbage truck of plastic is dumped into our oceans: this has to stop', *World Economic Forum*, [online], 27 October, Available from: https://www.weforum.org/agenda/2016/10/every-minute-one-garbage-truck-of-plastic-is-dumped-into-our-oceans/ [Accessed 18 December 2020].

Pennington, T. and Luther, D. (2014) *The Prepper's Blueprint*, np: CreateSpace Independent Publishing Platform.

Pereira, C. (2017) 'Get ready for zero waste week with these books', Book Riot, [online], Available from: https://bookriot.com/zero-waste-week-books/ [Accessed 18 May 2021].

Perkins, R., Berton, L. and Borulev, A. (2020) 'BP sells petrochemicals business to Ineos for $5 billion', *S&P Global*, [online], 29 June, Available from: https://www.spglobal.com/platts/en/market-insights/latest-news/petrochemicals/062920-bp-sells-petrochemicals-business-to-ineos-for-5-billion [Accessed 14 May 2021].

Petts, J. (1998) 'Trust and waste management information expectation versus observation', *Journal of Risk Research*, 1(4): 307–20.

Petts, J. (2001) 'Evaluating the effectiveness of deliberative processes: waste management case studies', *Journal of Environmental Planning and Management*, 44(2): 207–26.

Pijnenborg, R. and Vercruysse, L. (2008) 'Shifting concepts of the fetal–maternal interface: a historical perspective', *Placenta*, 22: S20–S25.

Pizzi, M., Fassan, M., Cimino, M., Zanardo, V. and Chiarelli, S. (2012) 'Realdo Colombo's De Re Anatomica: the renaissance origin of the term "placenta" and its historical background', *Placenta*, 33: 655–7.

Plastics Europe (2018) 'Plastics: the facts 2018; an analysis of European plastics production, demand and waste data', [online], Available from: https://www.plasticseurope.org/en/resources/publications/619-plastics-facts-2018 [Accessed 7 December 2020].

Pollans, L.B. (2017) 'Trapped in trash: "modes of governing" and barriers to transitioning to sustainable waste management', *Environment and Planning A: Economy and Space*, 49(10): 2300–23.

Popken, B. (2020) 'From disaster bunkers to dried food, survival supply sales are spiking', *NBC News Web*, [online], 13 March, Available from: https://www.nbcnews.com/business/consumer/disaster-bunkers-dried-food-survival-supply-sales-are-spiking-n1156921 [Accessed 7 December 2020].

Princen, T., Maniates, M. and Conca, K. (eds) (2002) *Confronting Consumption*, Cambridge, MA: MIT Press.

Rapoza, K. (2021) 'With Russia's help, China becomes plastics making power in pandemic', *Forbes*, [online], 28 January, Available from: https://www.forbes.com/sites/kenrapoza/2021/01/18/with-russias-help-china-becomes-plastics-making-power-in-pandemic/?sh=2a19299735a8 [Accessed 20 February 2021].

Reeves, H. (2001) 'A trail of refuse', *New York Times Magazine*, 18 February.

Reno, J. (2019) *Military Waste: The Unexpected Consequences of Permanent War Readiness*, Oakland: University of California Press.

Reuters, T. (2021) 'Biden ready to rejoin Paris Agreement, put the U.S. back on track with climate goals', *CBC/Radio-Canada*, [online], 20 January, Available from: https://www.cbc.ca/news/technology/biden-paris-agreement-climate-1.5880235 [Accessed 5 March 2021].

Robbins, P. and Moore, S.A. (2013) 'Ecological anxiety disorder: diagnosing the politics of the Anthropocene', *Cultural Geographies*, 20(1): 3–19.

Rocher, L. (2020) 'Waste, a matter of energy: a diachronic analysis (1992–2017) of waste-to-energy rationales', in N. Johansson and R. Ek (eds), *Perspectives on waste from the social sciences and humanities: Opening the Bin*, Newcastle upon Tyne: Cambridge Scholars Press, pp 98–117.

Rogers, H. (2006) *Gone Tomorrow: The Hidden Life of Garbage*, New York: New Press; Signature Book Services.

Rollinson, A. and Oladejo, J. (2020) 'Chemical recycling: status, sustainability, and environmental impacts', Global Alliance for Incinerator Alternatives.

Rosenfeld, J. (2013) 'Statistics on truck accident fatalities', [online], Available from: https://www.rosenfeldinjurylawyers.com/news/commercial-truck-fatality-statistics/ [Accessed 22 March 2017].

Roth, W.M. and Désautels, J. (2004) 'Educating for citizenship: reappraising the role of science education', *Canadian Journal of Science, Mathematics, and Technology Education*, 4(2): 149–68.

Roush, W. (2019) 'Wait, plastic can be good for the environment?', *Scientific American*, [online], 1 December, Available from: https://www.scientificamerican.com/article/wait-plastic-can-be-good-for-the-environment/ [Accessed 21 December 2020].

Rowe, R.K. (2012) 'Third Indian Geotechnical Society: Ferroco Terzaghi oration design and construction of barrier systems to minimize environmental impacts due to municipal solid waste leachate and gas', *Indian Geotechnical Journal*, 42(4): 223–56.

Royer, S-J., Ferrón, S., Wilson, S.T. and Karl, D.M. (2018) 'Production of methane and ethylene from plastic in the environment', *PLoS ONE*, 13(8): e0200574.

Salafia, C.M. and Vintzileos, A.M. (1990) 'Why all placentas should be examined by a pathologist in 1990', *American Journal of Obstetrics and Gynecology*, 163: 1282–93.

Samuel, G., Kerridge, I. and O'Brien, T. (2011) 'Ethical issues surrounding umbilical cord blood donation and banking', in N. Bhattacharya and P. Stubblefield (eds) *Regenerative Medicine Using Pregnancy-Specific Biological Substances*, London: Springer, pp 443–52.

Sarkodie, S.A. and Owusu, P.A. (2020) 'Impact of meteorological factors on COVID-19 pandemic: evidence from top 20 countries with confirmed cases', *Environmental Research*, 191: 1–7.

Saul, G. (2018) 'Environmentalists, what are we fighting for?', George Cedric Metcalf Charitable Foundation, October, pp 1–56.

Scalinci, S.Z., Scorolli, L., Corradetti, G., Domanico, D., Vingolo, E.M., Meduri, A., Bifani, M. and Siravo, D. (2011) 'Potential role of intravitreal human placental stem cell implants in inhibiting progression of diabetic retinopathy in type 2 diabetes: neuroprotective growth factors in the vitreous', *Clinical Ophthalmology*, 5: 691–6.

Schliesmann, P. (2011a) 'Reduce, reuse, revamp?', *Kingston Whig-Standard*, [online], 19 July, Available from: http://www.thewhig.com/2011/07/19/reduce-reuse-revamp [Accessed 3 September 2013].

Schliesmann, P. (2011b) 'Roots of mass recycling can be traced to Kitchener man', *Kingston Whig-Standard*, 18 July, 104.

Schliesmann, P. (2012) 'Special report: green challenge', *Kingston Whig-Standard*, [online], 12 March, Available from: http://www.thewhig.com/2012/03/12/special-report-green-challenge [Accessed 3 September 2013].

Schneider-Mayerson, M. (2013) 'Disaster movies and the "peak oil" movement: does popular culture encourage eco-apocalyptic beliefs in the United States?', *Journal for the Study of Religion, Nature and Culture*, 7(3): 289–314.

Shell (nd) 'Pennsylvania petrochemical complex', [online], Available from: https://www.shell.com/about-us/major-projects/pennsylvania-petrochemicals-complex.html [Accessed 9 March 2021].

Sheu, J. and Kuo, H. (2020) 'Dual speculative hoarding: a wholesaler–retailer channel behavioural phenomenon behind potential natural hazard threats', *International Journal of Disaster Risk Reduction*, 44: 1–10.

Silva, A.L.P., Prata, J.C., Walker, T.R., Duarte, A.C., Ouyang, W., Barcelo, D. and Rocha-Santos, T. (2020) 'Increased plastic pollution due to COVID-19 pandemic: challenges and recommendations', *Chemical Engineering Journal*, 405: 1–9.

Simpson, A. (2014) *Mohawk Interruptus: Political Life across the Borders of Settler States*, Durham, NC: Duke University Press.

Skill, K. (2008) '(Re)Creating ecological action space: householders' activities for sustainable development in Swedent'. PhD thesis, Linköping University, Sweden.

Smil, V. (2016) *Energy Transitions: Global and National Perspectives, Second Edition*. Westport, CT: Praeger.

Smith, R.J. (1982) 'The risks of living near Love Canal', *Science*, 217: 808–9, 811.

Sohrabi, C., Alsafi Z., O'Neill, N., Khan, M., Kerwan, A., Al-Jabir, A., Iosifidis, C. and Agha, R. (2020) 'World Heath Organization declares global emergency: a review of the 2019 novel coronavirus (COVID-19)', *International Journal of Surgery*, 76: 71–6.

Solomon, B.D., Andrén, M. and Strandberg, U. (2012) 'Three decades of social science research on high-level nuclear waste: achievements and future challenges', *Risk, Hazard, and Crisis in Public Policy*, 1(4): 13–47.

Statistics Canada (2012) 'Human activity and the environment', [online], Available from: https://www150.statcan.gc.ca/n1/pub/16-201-x/16-201-x2012000-eng.htm [Accessed 14 May 2021].

Statistics Canada (2020) 'Canadian consumers prepare for COVID-19', [online], Available from: https://www150.statcan.gc.ca/n1/en/pub/62f0014m/62f0014m2020004-eng.pdf?st=tk2_Wyh_[Accessed 9 March 2021].

Staub, C. (2021) 'Surprise rise: paper and plastic exports up to start 2021', *Resource Recycling*, [online], 9 March, Available from: https://resource-recycling.com/recycling/2021/03/09/surprise-rise-paper-and-plastic-exports-up-to-start-2021/ [Accessed 9 March 2021].

Stein, M. (2008) *When Technology Fails: A Manual for Self-Reliance, Sustainability and Surviving the Long Emergency*, White River Junction: Chelsea Green Publishing.

Stengers, I. (2005) 'The cosmopolitical proposal', in B. Latour and P. Weibel (eds), *Making Things Public*, Cambridge, MA: MIT Press, pp 994–1003.

Stopline3.org (nd) 'Stop the Line 3 pipeline', [online], Available from: https://www.stopline3.org/ [Accessed 18 May 2021].

Stow, J.P., Sova, J. and Reimer, K.J. (2005) 'The relative influence of distant and local (DEW-line) PCB sources in the Canadian Arctic', *Science of the Total Environment*, 342(1–3): 107–18.

Strasser, S. (1999) *Waste and Want: A Social History of Trash*, New York: Henry Holt and Company.

Sullivan, P. (2020) 'Wealthy fliers worried about coronavirus turn to private jet service', *New York Times*, [online], 30 May, Available from: https://www.nytimes.com/2020/05/30/your-money/coronavirus-private-jets.html [Accessed 9 March 2021].

Supran, G. and Oreskes, N. (2020) Reply to Comment on 'Assessing ExxonMobil's climate change communications (1977–2014)' Supran and Oreskes (2017) *Environment Research Letters*, 15(11): 118002.

Switzer, J. (2008) 'Having a little enviro-guilt can be a good thing', *Kingston Whig-Standard*, 16 July, 1–2.

Szasz, A. (2007) *Shopping Our Way to Safety: How We Changed from Protecting the Environment to Protecting Ourselves*, Minneapolis: University of Minnesota Press.

Tactical.com (nd) 'A complete beginner's guide to prepping', [online], Available from: https://www.tactical.com/complete-beginners-guide-to-prepping/ [Accessed 26 October 2020].

Takai, Y., Tsutsumi, O., Ikezuki, Y., Hiroi, H., Osuga, Y., Momoeda, M., Yano, T. and Taketani, Y. (2000) 'Estrogen recipient-mediated effects of a Xenoestrogen, Bisphenol A on preimplantation mouse embryos', *Biochemistry and Biophysical Research Communications*, 270(3): 918–21.

Tellefsen, C.H. and Vogt, C. (2011) 'How important is placental examination in cases of perinatal deaths?', *Pediatric and Developmental Pathology*, 14: 99–104.

Thompson, E.P. (1966) *The Making of the English Working Class*, New York: Vintage Books.

Thompson, J. and Anthony, H. (2008) 'The health effects of waste incineration', *Fourth Report of the British Society for Ecological Medicine* (2nd edn), [online], Available at: http://www.ecomed.org.uk/content/IncineratorReport_v3.pdf [Accessed 2 October 2012].

Thomson, V.E. (2009) *Garbage in, Garbage Out: Solving the Problems with Long-Distance Trash Transport*, Charlottesville: University of Virginia Press.

Tibbetts, J. (2013) 'Garbage collection is "one of the most hazardous jobs"', *CMAJ*, 185(7): E284.

Todd, Z. (2016) 'An indigenous feminist's take on the ontological turn: "ontology" is just another word for colonialism', *Journal of Historical Sociology*, 29(1): 4–22.

Toh, M. (2020) ' "It's crazy": panic buying forces stores to limit purchases of toilet paper and masks', *CNN*, [online], 6 March, Available from: https://www.cnn.com/2020/03/06/business/coronavirus-global-panic-buying-toilet-paper/index.html [Accessed 26 October 2020].

Tompkins, J. (2019) 'Just one word: plasticurious', Canadian Centre for Policy Alternatives, Available from: https://monitormag.ca/articles/just-one-word-plasticurious/ [Accessed 5 January 2020].

Toomey, C. (2008) 'It's easier than you think to put a lid on coffee cup waste', *Kingston Whig-Standard*, 17 November, 1–2.

Tourism Montréal (nd) [online], Available from: https://www.mtl.org/en/what-to-do/activities/frederic-back-park-montreal [Accessed 11 May 2021].

Tuhiwai Smith, L. (2012) *Decolonizing Methodologies: Research and Indigenous Peoples*, 2nd edn, London: Zed Books.

UNCTAD (2020) 'Growing plastic pollution in wake of COVID-19: how trade policy can help', [online], 27 July, Available from: https://unctad.org/news/growing-plastic-pollution-wake-covid-19-how-trade-policy-can-help [Accessed 6 May 2021].

UNESCO International School of Science for Peace (1998) 'Nuclear disarmament, safe disposal of nuclear materials or new weapons developments? Where are the national laboratories going?', Edited by P.C. Ramusino, G. Gherardi, A. Lantieri V. Kouzminov, M. Martellini and R. Santesso, Venice: UNESCO Venice Office.

UNHCR (2016) 'UNHCR: Global Trends, Forced Displacement in 2016', [online], Available from: http://www.unhcr.org/5943e8a34.pdf

US Chamber of Commerce (2020) 'Multi-association letter to Minister Mary Ng on CEPA plastics issue', [online], Available from: https://www.uschamber.com/comment/multi-association-letter-minister-mary-ng-cepa-plastics-issue [Accessed 24 May 2021].

US Department of Energy (2006) 'Recycling paper and glass', [online], Available from: http://www.eia.doe.gov/kids/energyfacts/saving/recycling/solidwaste/paperandglass.html [Accessed 15 September 2012].

US Energy Information Administration (2020) 'Frequently Asked Questions', [online], Available from: https://www.eia.gov/tools/faqs/faq.php?id=34&t=6

Van Ewijk, S. and Stegemann, J. (2014) 'Limitations of the waste hierarchy for achieving absolute reductions in material throughput', *Journal of Cleaner Production*, 132: 122–8.

Van Oostdam, J., Donaldson, S.G., Feeley, M., Arnold, D., Ayotte, P., Bondy, G., Chan, L, Dewaily, É., Furgal, C.M., Kuhnlein, H., Loring, E., Muckle, G., Myles, E., Receveur, O., Tracy, B., Gill, U. and Kalhok, S. (2005) 'Human health implications of environmental contaminants in Arctic Canada: a review', *Science of the Total Environment*, 351–2: 165–246.

Van de Poel, I. (2008) 'The bugs eat the waste: what else is there to know? Changing professional hegemony in the design of sewage treatment plants', *Social Studies of Science*, 38(4): 605–34.

van Wyck, P. (2005) *Signs of Danger: Waste, Trauma, and Nuclear Threat*, Minneapolis: University of Minnesota Press.

Vendries, J., Sauer, B., Hawkins, T., Allaway, D., Canepa, P., Rivin, J. and Mistry, M. (2020) 'The significance of environmental attributes as indicators of the life cycle environmental impacts of packaging and food service ware', *Environmental Science and Technology*, 54: 5356–64.

Vigdor, N. (2020) 'A hoarder's huge stockpile of masks and gloves will now go to doctors and nurses, F.B.I. says', *New York Times*, 2 April, Available from: https://www.nytimes.com/2020/04/02/nyregion/brooklyn-coronavirus-price-gouging.html [Accessed 24 May 2021].

Villareal, D. (2020) 'Ukrainian church leader who blamed COVID-19 on gay marriage tests positive for virus', *NBC News*, [online], 8 September, Available from: https://www.nbcnews.com/feature/nbc-out/ukrainian-church-leader-who-blamed-covid-19-gay-marriage-tests-n1239528 [Accessed 6 May 2021].

Wagner, S. (2019) 'Can you own a private jet if you care about climate change?', *BNN Bloomberg*, [online], 22 December, Available from: https://www.bnnbloomberg.ca/can-you-own-a-private-jet-if-you-care-about-climate-change-1.1365986 [Accessed 6 May 2021].

Wang, M., Yang, Y., Yang, D., Luo, F., Liang, W., Guo, S. and Xu, J. (2008) 'The immunomodulatory activity of human umbilical cord blood-derived mesenchymal stem cells in vitro', *Immunology*, 126: 220–32.

Wapner, P. (1996) 'Toward a meaningful ecological politics', *Tikkun*, 11(3): 21–2.

Waste360.com (2021) 'P&G and its brands, like Pantene, Gillette, Ariel, Fairy and Oral B are reducing virgin plastic', [online], 8 March, Available from: https://www.waste360.com/recycling/pg-and-its-brands-pantene-gillette-ariel-fairy-and-oral-b-are-reducing-virgin-plastic [Accessed 13 March 2021].

Watt-Cloutier, S. (2015) *The Right to Be Cold: One Woman's Story of Protecting Her Culture, the Arctic and the Whole Planet*, Toronto: Penguin.

Weisgall, J. (1994) *Operation Crossroads: The Atomic Tests at Bikini Atoll*, Annapolis: Naval Institute Press.

Weyler, R. (2019) 'It's a waste world', *Greenpeace*, [online], 20 July, Available from: https://www.greenpeace.org/international/story/23747/its-a-waste-world/ [Accessed 19 May 2021].

Wheeler, K. and Glucksmann, M. (2015) '"It's kind of saving them a job isn't it?" The consumption work of household recycling', *Sociological Review*, 63(3): 551–69.

Williams, A. and Bromwich, J.E. (2020) 'The rich are preparing for the coronavirus differently', *New York Times*, [online], 5 March, Available from: https://www.nytimes.com/2020/03/05/style/the-rich-are-preparing-for-coronavirus-differently.html [Accessed 7 October 2020].

Wilson, D.C., Ljiljana, R., Prasad, M., Soos, R., Rogero, A.C., Velis, C. et al (2009) 'Global waste management outlook', Vienna: United Nations Environment Programme/International Solid Waste Association.

Wilson, D.C. and Velis, C.A. (2015) 'Waste management: still a global challenge in the 21st century; an evidence-based call for action', *Waste Management & Research*, 33(12):1049–51.

Winter, L. (2020) 'Chinese officials blame US Army for coronavirus', *The Scientist*, [online], 13 March, Available from: https://www.the-scientist.com/news-opinion/chinese-officials-blame-us-army-for-coronavirus-67267 [Accessed 6 May 2021].

Wolff, P., Tanaka, A.M., Chenker, E., Cabrera-Crespo, J., Raw, I. and Ho, P.L. (1996) 'Purification of fibroblast growth factor-2 from human placenta using tri(n-butyl)phosphate and sodium cholate', *Biochimie*, 78: 190–4.

Woodroof, N. (2021) 'SIBUR and Sinopec create JV at Amur Gas Chemical Complex', *Hydrocarbon Engineering*, [online], Available from: https://www. hydrocarbonengineering.com/petrochemicals/04012021/sibur-and-sino pec-create-jv-at-amur-gas-chemical-complex/ [Accessed 10 February 2021].

World Economic Forum, Ellen MacArthur Foundation and McKinsey & Company (2016) *The New Plastics Economy — Rethinking the future of plastics*, [online], Available from: http://www.ellenmacarthurfoundation. org/publications).

World Health Organization (WHO) (nd) 'WHO coronavirus (COVID-19) dashboard', [online], Available from: https://covid19.who.int/ [Accessed 21 May 2021].

Wynne, B. (1987) *Risk Management and Hazardous Waste: Implementation and Dialectics of Credibility*, Heidelberg: SpringerVerlag.

Yang, S. and Furedy, C. (1993) 'Recovery of wastes for recycling in Beijing', *Environmental Conservation*, 20: 79–82.

Yen, B.L., Huang, H-I., Chien, C-C., Jui, H-Y., Ko, B-S., Yao, M., Shun, C-T., Yen, M-L., Lee, M-C. and Chen, Y-C. (2005) 'Isolation of multipotent cells from human term placenta', *Stem Cells*, 23: 3–9.

Yoshizawa, R.S. (2013) 'Public perspectives on the utilization of human placentas in scientific research and medicine', *Placenta*, 34: 9–13.

Yoshizawa, R.S. (2014) 'Placentations: agential realism and the science of afterbirths', PhD Thesis, Queen's University. Available from: https://qsp ace.library.queensu.ca/handle/1974/12408.

Yoshizawa, R.S. (2016) 'Fetal–maternal intra-action: politics of new placental biologies', *Body & Society*, 22(4): 79–105.

Young, S.M. and Benyshek, D.C. (2010) 'In search of human placentophagy: a cross-cultural survey of human placenta consumption, disposal practices, and cultural beliefs', *Ecology of Food and Nutrition*, 49: 467–84.

Yu, S.J., Soncini, M., Kaneko, Y., Hess, D.C., Parolini, O. and Borlongan, C.V. (2009) 'Amnion: a potent graft source for cell therapy in stroke', *Cell Transplantation*, 18: 111–18.

Zhang, X., Mitsuru, A., Igura, K., Takahashi, K., Ichinose, S., Yamaguchi, S. and Takahashi, T.A. (2006) 'Mesenchymal progenitor cells derived from chorionic villi of human placenta for cartilage tissue engineering', *Biochemical and Biophysical Research Communications*, 340: 944–52.

Zhu, X. (2021) 'Plastic is part of the carbon cycle and needs to be included in climate calculations', *The Canadian Press*, [online], 1 March, Available from: https://nationalpost.com/pmn/news-pmn/plastic-is-part-of-the-car bon-cycle-and-needs-to-be-included-in-climate-calculations [Accessed 9 March 2021].

Index

References to figures appear in *italic* type. References to endnotes show both the
page number and the note number (109n1)